Biomaterial Mech

Biomaterial Mechanics

Edited by
Heather N. Hayenga and
Helim Aranda-Espinoza

CRC Press
Taylor & Francis Group
Boca Raton London New York

CRC Press is an imprint of the
Taylor & Francis Group, an **informa** business

CRC Press
Taylor & Francis Group
6000 Broken Sound Parkway NW, Suite 300
Boca Raton, FL 33487-2742

First issued in paperback 2019

ISBN-13: 978-1-4987-5268-8 (hbk)
ISBN-13: 978-0-367-87585-5 (pbk)

Library of Congress Cataloging-in-Publication Data

Names: Hayenga, Heather N., editor. I Aranda-Espinoza, Helim, editor.
Title: Biomaterial mechanics / [edited by] Heather N. Hayenga and Helim
Aranda-Espinoza.
Description: Boca Raton : CRC Press/Taylor & Francis, 2017. I Includes
bibliographical references and index.
Identifiers: LCCN 2017011216 I ISBN 9781498752688 (hardback : alk. paper)
Subjects: I MESH: Biocompatible Materials I Biomimetic Materials I
Biomechanical Phenomena I Bioengineering--methods I Computer Simulation
Classification: LCC R857.M3 I NLM QT 37 I DDC 610.28--dc23
LC record available at https://lccn.loc.gov/2017011216

Professor Heather N. Hayenga dedicates this book in loving memory to her generous and selfless father, Calvin Godfrey Hayenga, III.

Professor Helim Aranda-Espinoza dedicates this book to Joanna, Amaya, and Logan.

Table of Contents

Preface

This book, *Biomaterial Mechanics*, comes in a timely manner as the biomedical field is realizing the importance mechanics plays in obtaining a successful healthcare outcome. Thus when asked by the publishers if we could coedit a book emphasizing the role of mechanics in biomaterials, we gladly accepted. This book provides an overview of the fundamental mechanical principles of biomaterials and the types of biomaterials used in healthcare. Moreover, an overview of elegant computational modeling approaches is presented with the applicable insights toward the refinement of biomaterial mechanics and biocompatibility.

In the wake of new technological advancements, biomedical materials have evolved in complexity and have played a crucial role in treating health ailments. The success of the biomaterial relies on multifaceted mechanical and biological interactions between the material and the host. Traditionally, materials were considered nearly homogeneous and assumptions to simplify material properties were appropriate, including assumptions that the materials only undergo small strains and the change in strain is linearly proportional to the change in stress. However, these assumptions are typically not appropriate when considering biological tissues that may undergo large strains, directional dependencies, or nonlinear stress–strain behavior. The first part of this book reviews mechanical principles of biomaterials, including how to appropriately test and calculate the mechanical properties of soft biological materials.

The next part of this book overviews the evolution of engineered biomaterials and implantable devices since 1950 when the field of biomaterials started to blossom. The increasing degree of sophistication has enabled biomaterials today to incorporate biological active components and dynamic behavior giving them crucial roles in injury repair, diagnostics, biological screening, drug delivery, and tissue engineering applications. There are many novel biomaterials on the horizon, including smart materials, shape memory polymers, 3D-printed biomaterials, and nanomaterials that have the potential to revolutionize healthcare in the future. This part specifically reviews properties of metallic, polymeric, nano, biological, and cancer-related biomaterials highlighting the key mechanical discoveries that will enable the eventual design, synthesis, characterization, and implantation of the optimal biomimetic material to treat a particular disease or substitute a tissue or entire organ.

In addition to the advanced manufacturing processes, computational modeling has allowed for a better understanding of how physical properties affect biological performance, as well as the interplay between various physio-mechano-chemical properties. The third part of this book highlights how computational biomechanical models can advance the field of device design. Modeling can enable time- and cost-efficient evaluations of fundamental hypotheses and thus reduce the experimental search space, and aid investigators in moving away from a purely empirical, trial-and-error approach toward rational design. In this part, the authors give concrete examples, including how computation modeling was used to determine optimal scaffold geometry, stiffness, degradation rates, and production rates of vascular grafts. Moreover, modeling can be used on a smaller scale to determine the molecular physicochemical interactions in biological systems and between biomaterials. More generally, a method called finite element analysis can be used to calculate stresses within biomaterials and tissues and thus proves a useful tool in evaluating the response of host tissue to an implanted biomaterial. In this part, the authors present the general principles of FEA and highlight how it is useful in evaluating the design and properties of cardiovascular stents. Overall, inclusive models that apply principles from engineering, material science, and biology will help develop and understand the mechanics of new materials better.

In summary, new processing techniques have enabled the fabrication of biomaterials that are more biomimetic than ever before. It can even be argued that some biomaterials have surpassed purely biomimetic materials. Moreover, advanced computational modeling has allowed for a better understanding of how physical properties affect biological performance, as well as the interplay between various physio-mechano-chemical properties. The ability to multiplex physical and chemical design parameters among single and multiple materials will help ensure compatibility and optimal performance of biomedical devices. It is an exciting time to explore the various parameters that control the physio-mechano-chemical properties of biomaterials. This book will help provide the background knowledge needed to pursue such endeavors.

The authors of this book are leading experts of their respective content. Each chapter is written independently of one another and can therefore be read individually and not necessarily in sequence. However, the logical progression of the book will provide interested readers with a comprehensive knowledge on how to test and characterize the mechanical properties of biomaterials, the types and mechanical properties of biomaterials used in the biomedical field, the advantage and insights obtained through computationally modeling the mechanical and biological interactions of biomaterials, and finally hypothesize as to where the field of novel biomaterials is headed in the future. Personally, as editors, we have learned a lot on each subject and expect the reader will also learn many new facets of the evolving field of biomaterials. Please enjoy your copy of *Biomaterial Mechanics*!

Heather N. Hayenga
University of Texas at Dallas
Richardson, Texas

Helim Aranda-Espinoza
University of Maryland
College Park, Maryland

Editors

Dr. Heather N. Hayenga is an assistant professor in the bioengineering department at the University of Texas at Dallas. She earned her BS in biomedical engineering on June 2005 from the University of California, Davis, and her PhD in biomedical engineering in 2011 from Texas A&M University. Dr. Hayenga then joined the University of Maryland as a postdoctoral fellow researching cellular biophysical phenomenon. To date, she has published 14 peer-reviewed manuscripts in the fields of biomechanics, mechanobiology, vascular biology, and immunology; has given more than 30 conference and invited talks; serves as a faculty advisor for the Biomedical Engineering Society (BMES) and the Society of Women Engineers (SWE); and serves as a reviewer for many grant and journal organizations. After joining UT Dallas in 2013, Dr. Hayenga developed and principals the Vascular Mechanobiology Laboratory. Her primary research interests include developing predictive growth and modeling tools of the vasculature, investigating the mechanobiology of arterial cells, improving drug delivery through the blood–brain barrier, and refining cardiovascular biomaterials during stenting.

Dr. Helim Aranda-Espinoza is an associate professor in the Fischell Department of Bioengineering at the University of Maryland, College Park. Aranda-Espinoza leads the Cell Biophysics Laboratory. To date, he has published 42 peer-reviewed manuscripts in the fields of soft matter, leukocyte biophysics, lipid dynamics, neuroengineering, and, recently, cancer metastasis. He has received prestigious awards such as the NSF Career Award and the Human Frontier Science Program Award. His primary research interests include understanding how the mechanical environment dictates cell functions. This is accomplished through the application of theoretical and experimental physics and engineering to describe cell mechanics and problems encountered in biological systems.

Contributors

Katrina Adlerz
Fischell Department of Bioengineering
University of Maryland
College Park, Maryland

Shant Aghyarian
Department of Bioengineering
University of Texas at Dallas
Richardson, Texas

Said Aranda
Instituto de Física
Universidad Autónoma de San Luis
    Potosí
San Luis Potosí, Mexico

Helim Aranda-Espinoza
Fischell Department of Bioengineering
University of Maryland
College Park, Maryland

Christopher K. Breuer
Tissue Engineering Program and
    Surgical Research
Nationwide Children's Hospital
Columbus, Ohio

Matthew Di Prima
Center for Devices and Radiological
    Health
U.S. Food and Drug Administration
Silver Spring, Maryland

John F. Eberth
Cell Biology and Anatomy Department
and
Biomedical Engineering Program
University of South Carolina
Columbia, South Carolina

John P. Fisher
Fischell Department of Bioengineering
University of Maryland
College Park, Maryland

Sai J. Ganesan
Fischell Department of Bioengineering
University of Maryland
College Park, Maryland

Edna George
Department of Biosciences and
    Bioengineering
Indian Institute of Technology Bombay
Mumbai, India

Izabelle M. Gindri
Department of Bioengineering
University of Texas at Dallas
Richardson, Texas

Heather N. Hayenga
Department of Bioengineering
University of Texas at Dallas
Richardson, Texas

Kim L. Hayenga
Biomedical Engineer
San Mateo, California

Jay D. Humphrey
Department of Biomedical Engineering
Yale University
New Haven, Connecticut

Ramak Khosravi
Department of Biomedical Engineering
Yale University
New Haven, Connecticut

Silvina Matysiak
Fischell Department of
    Bioengineering
University of Maryland
College Park, Maryland

Clark A. Meyer
Department of Bioengineering
University of Texas at Dallas
Richardson, Texas

Kristin S. Miller
Department of Biomedical Engineering
Tulane University
New Orleans, Louisiana

Jesse K. Placone
Department of Bioengineering
University of California, San Diego
San Diego, California

Mildred Quintana
Instituto de Física
Universidad Autónoma de San Luis
  Potosí
San Luis Potosí, Mexico

Radu Reit
Department of Bioengineering
University of Texas at Dallas
Richardson, Texas

Danieli C. Rodrigues
Department of Bioengineering
University of Texas at Dallas
Richardson, Texas

Lucas Rodriguez
Department of Bioengineering
University of Texas at Dallas
Richardson, Texas

Shamik Sen
Department of Biosciences and
  Bioengineering
Indian Institute of Technology Bombay
Mumbai, India

Tarek Shazly
Mechanical Engineering Department
and
Biomedical Engineering Program
University of South Carolina
Columbia, South Carolina

Sathyanarayanan Sridhar
Department of Bioengineering
University of Texas at Dallas
Richardson, Texas

Walter E. Voit
Department of Mechanical
  Engineering and Materials Science
  and Engineering
University of Texas at Dallas
Richardson, Texas

PART I
PRINCIPLES OF BIOMATERIAL MECHANICS

1 Overview of Mechanical Behavior of Materials

Radu Reit, Matthew Di Prima, and Walter E. Voit

CONTENTS

1.1 INTRODUCTION: MECHANICS OF MATERIALS

The understanding of how a material behaves under load is critical for designing synthetic materials compatible with biological systems and for understanding and interfacing with structural biological materials. When given a set of operating parameters, engineers must pick the appropriate material for the task at hand to ensure the proper operation of their design. For scientists and clinicians, an effective understanding of load, fatigue, anisotropy, and the general frequency dependence of mechanical properties is useful for making advances in healthcare, patient treatment, and biomedicine. In the case of biomaterials, engineered medical devices constantly interact with a variety of materials that range in stiffness and fragility, such as soft brain tissue or hard bone. To engineer new biological materials, bio-inspired synthetic materials, and purely synthetic materials used in medical devices, or improve existing materials, an understanding of the mechanical interaction between biotic and abiotic interfaces is critical to ensure durability, longevity, and minimization of immunological responses. In this first chapter, the reader is introduced to the basics of how materials behave under a mechanical load, how to compare among materials given available reference values, and to approaches on how to design appropriate experiments to ascertain materials' data when they are not available and the materials of interest (synthetic or biological) are difficult to process into standard specimen geometries and, consequently, cannot always follow standardized testing protocols.

1.2 MECHANICAL BEHAVIOR OF MATERIALS

To fully understand the response of a material to an applied load, it is important to understand its fundamental atomic structure. Critical to this task is the knowledge of the various bonding types in which a material can participate (ionic, covalent, metallic, etc.) as well as whether the material exhibits a crystalline structure once bonded (face-centered cubic, body-centered cubic, hexagonal close-packed, etc.), is amorphous, or is a complex composite with spatially varying microstructures. As detailed knowledge of atomic structure can cover an entire multi-chapter book by itself, the reader is referred to a detailed text from Valim, among many others, which provides a nice primer on interatomic bonding in solids [6]. Therefore, the remainder of the text will guide the reader toward understanding the fundamental macroscopic properties of materials under load.

1.2.1 Elastic Response of Materials

When a material is studied at the macroscopic level, the bulk of information regarding its mechanical behavior is determined by its elastic response under load. A simple analogy for this is the behavior of an ideal Hookean spring in compression or tension: given the application of a fixed load F (usually measured in Newtons [N]), the spring will deform a given distance x (meters [m]), modified by a spring constant k (N/m), according to Equation 1.1:

$$F = -kx \qquad (1.1)$$

Given the cross-sectional area for the applied load and the ability to accurately measure the deformation of the samples, F can be equated into a stress, $\sigma = F/A_o$, where A_o is the initial cross-sectional area of the sample, and x can be equated to the strain, $\varepsilon(\Delta x/x_i)$. Thus, the Hookean formula in Equation 1.1 is rearranged into Equation 1.2:

$$\sigma = E\varepsilon \qquad (1.2)$$

where E is the Young's modulus (also known as the elastic modulus) of the material, a coefficient linearly relating stress to strain values. A stiff material can be thought of as a spring with a large spring constant, and will therefore have a correspondingly large E. Similarly, a soft material has a small spring constant and a similarly low E. These generalizations are shown in Figure 1.1, where the Hookean spring behavior is shown in Figure 1.1a and the corresponding stress versus strain behavior is plotted for a stiff and soft material in Figure 1.1b. Unfortunately, the elastic response of a material is only a small portion of its

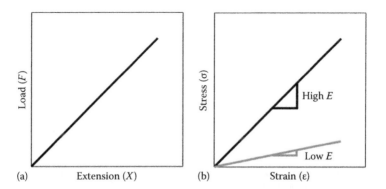

Figure 1.1 (a) Load versus extension, with the transformation to (b) stress versus strain and the representation of Young's modulus, E.

behavior during deformation. To further understand the behavior of a material under load, the entirety of the stress–strain response has to be considered, well outside of the initial so-called linear elastic regime, the region in which Equation 1.2 is valid for calculating E.

1.2.2 Stress–Strain Response

Often, mechanical characterization is used to guide the design of new materials, and focuses on the determination of the Young's modulus of a material in order to rapidly compare different materials and pick the appropriate material for a given application. To determine this modulus experimentally, the stress–strain behavior of the material must be analyzed first. Figure 1.2a demonstrates a representative stress–strain schematic for a material tested at a constant temperature. Given the stress applied σ, a strain ε is experimentally measured until the failure of the material (a break in the specimen). Using this methodology, a few mechanically distinct regions become apparent in a typical stress–strain curve.

First, the elastic region is shown as encompassed by the linear portion of the representative stress–strain graphic. In this elastic regime, the material behaves

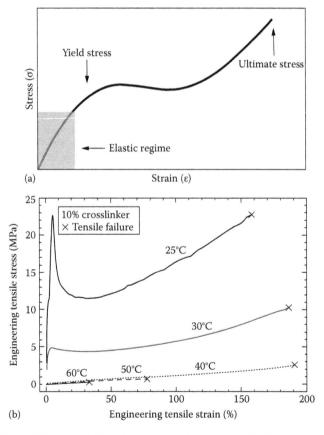

Figure 1.2 (a) Representative stress vs. strain curve, with (b) example curves for acrylic shape memory polymers at various loading temperatures. (Reproduced from Yakacki, C.M. et al., *Adv. Eng. Mater.*, 10, 112, 2008. With permission.)

as an ideal spring with the same behavior described by Equation 1.2. If fact, the Young's modulus (E) is calculated as the slope ($\Delta\sigma/\Delta\varepsilon$) during this initial portion of the stress–strain curve during a test where the load is extending or compressing the sample. The linear elastic regime is critical to understanding material operating parameters as it describes the range over which a sample will behave as the ideal Hookean spring, relieving strain in the material after release of the applied stress. As predicted by Equation 1.2, materials with a high E exhibit low strain and those with a low E exhibit high strain under the same loading. Part of the quest of materials scientists is to find, process, mimic, or design materials with increased stress *and* strain capacities, which leads to tougher materials. One measure of materials toughness is the integrated area underneath a stress–strain curve, or the amount of energy a material can absorb either per linear distance, per surface area, or per volume. Ceramics and stiff plastics are typically high-strength materials (can sustain high stress without failing), but are often brittle and have low toughness. On the other hand, soft polymers (elastomers) and tissues typically show increased strain capacity at the expense of lower modulus. Engineered metals and high-strength, but ductile, polymers (nylon, Kevlar, etc.) are among the class of tough materials. Many biological materials (cartilage, connective tissue, etc.) are incredibly tough when tested at very specific conditions (temperature, moisture, pH) but outside of this range fail miserably as structural materials. Thus, as we will explain in this chapter, the properties of materials are intimately tied to the properties of the surrounding environment, the amount of load applied, and the frequency of the applied load. Case studies will be shown in Section 1.3 about how different biomaterials (synthetic and natural) perform under varying loading regimes.

The next regime we present is that of plastic deformation in a sample, or irreversible deformation outside of the linear elastic regime. In this plastic section of the stress–strain curve, defined by a knee after the elastic region, the material does not exhibit ideal Hookean spring behavior found in the elastic regime. Instead, permanent changes in the microstructure caused by strain past the yield stress (σ_y) shown in Figure 1.2a cause the material to undergo an unrecoverable or only partially recoverable strain. That is to say that given the release of all stress applied to the material, the sample will not return to its initial configuration. While the plastic region is not used in calculating the modulus of a material, it is of crucial importance in determining how much energy the material can absorb before breaking (again known as the toughness of the sample). Figure 1.3a demonstrates two types of materials: a material with a short plastic region, known as a brittle material, and a material with a large plastic region, known as a ductile material. While also not a guaranteed predictor of material behavior, in general, substances such as metals, ceramics, and plastics that are covalently linked between chains will behave as brittle solids, while tissue and plastics without covalent linkages between chains are more ductile in nature. An example stress–strain curve for a blend of PP (polypropylene) and SBR (styrene–butadiene rubber) is shown by Jang et al. in Figure 1.3b, which depicts how the stress–strain behavior of the polymer can show brittle behavior at low temperatures and high strain rates (curve 10: $T = -17°C$, $\Delta\varepsilon/\Delta t = 20$ in./min) or ductile behavior at high temperatures and low strain rates (curve 1: $T = 23°C$, $\Delta\varepsilon/\Delta t = 0.005$ in./min) [3]. This time–temperature dependence of material stress–strain behavior is further explored in Sections 1.2.5 and 1.4.

1.2.3 Fracture Behavior

Material failure is typically defined when the test sample fractures during mechanical loading. In the case of brittle materials, the cross-sectional area at the fracture plane is very close in size and value to the initial cross-sectional area (also

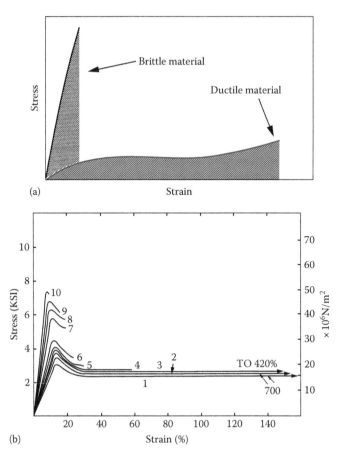

Figure 1.3 (a) Representative integrations for calculating the toughness of a brittle or ductile material. (b) Brittle–ductile transition in a polymer blend of polypropylene and styrene–butadiene rubber. (Reproduced from Jnag, B.Z. et al., *J. Appl. Poly. Sci.*, 29, 3409, 1984. With permission.)

known as the gauge area) measured for the specimen. An important phenomenon of plastic deformation is that it becomes important to understand the changing specimen cross-sectional area and note that the actual stress (true stress: σ_T) and the measured stress (engineering stress) differ in value due to a changing cross-sectional area of the material. This also reflects in a modified strain (true strain: ε_T) and leads to sample behavior governed by the following equations:

$$\sigma_T = \frac{F}{A} \tag{1.3}$$

$$\varepsilon_T = \ln\frac{x}{x_i} \tag{1.4}$$

where A is the actual cross-sectional area of the specimen during the application of a load, and the true strain is redefined by the natural log of the current dimension (x) divided by the initial dimension (x_i). What is important to note

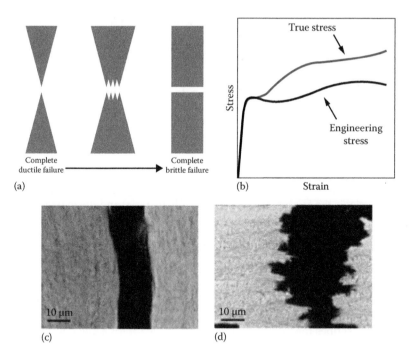

Figure 1.4   (a) Simplified fracture morphologies for ductile versus brittle failures. (b) Representative true versus engineering stress behavior. (c) Brittle and (d) semiductile behavior of femoral bone depending on load direction (longitudinal and circumferential, respectively). (Reproduced from Peterlik, H. et al., *Nat. Mater.*, 5, 52, 2006. With permission.)

is that barring any break in the sample, the true stress on the material does not decrease in value with any incremental increase in strain, unlike the engineering stress that can show decreases before rupture of the specimen. Although the true stress yields a more correct quantitative assumption of the material properties, for practical applications, the engineering stress–strain curve gives a more accurate approximation of the behavior under load. Therefore, by far the most common stress–strain data presented when testing materials is the engineering stress–strain, as this determines the behavior of the material under realistic loads without concern for change in cross-sectional area. Figure 1.4 shows the practical considerations of this phenomenon with exemplary fracture morphologies of various elongations in materials (Figure 1.4a), as well as what the corresponding stress–strain graphs would look like for the ductile failure of materials in terms of engineering and true stress–strain behavior (Figure 1.4b). The phenomenon observed in Figure 1.4a is known as necking of the sample, defined as a reduction in cross-sectional area of the specimen being tested. The various morphologies seen in this figure show an extremely ductile material (severe necking to a single-point failure), an intermediate material with minimal necking, and an observable fracture plane, and finally an ideal brittle material with no necking and a clear fracture plane. This phenomenon was shown experimentally in bone by Peterlik et al. in 2006, where a longitudinal fracture (along the length of the bone) in femoral bone samples leads to a clean fracture plane representative of brittle failure (Figure 1.4c) and a circumferential fracture (along the radius of the bone) leads to a semiductile fracture behavior (Figure 1.4d) [4]. These variable fracture mechanics in the same material hint at the concept of anisotropy, discussed later in Section 1.2.6.

A final critical metric derived from the stress–strain behavior of a material is its ultimate strength (referred to as the ultimate tensile strength [UTS] when the stress–strain data are collected in tension). The ultimate strength of a material refers to the maximum stress that a material withstood under mechanical loading; again, when looking at engineering versus true stress–strain curves, this metric will be quantitatively different. However, as engineering stress/ strain is the principal focus of this text (as well as most practical mechanical applications), the ultimate strength of a material is measured as the maximum engineering stress withstood by the sample. In other words, the ultimate strength of the material is the fracture strength of a brittle material and the maximum stress experienced by a ductile material before necking begins. Unfortunately, the ultimate strength of a material does not give the complete representation of a material's behavior under cyclical loading.

1.2.4 Fatigue

Another important material behavior is the performance under cyclical loading, known as the fatigue characteristics of a material. To quantify this behavior in a material, one must look at the endurance limit of a material, or the S–N curve that describes the maximum stress amplitude (S) for a given number of loading cycles (N). As shown in Figure 1.5a, most materials will follow an exponential decay that eventually reaches a nonzero stress under which the material can be loaded indefinitely (black line). This asymptote is known as the endurance limit of a material and describes an operating load under which the material is stable indefinitely, notwithstanding damage from other external factors. The important fact to note, however, is that not all materials will show an endurance limit; that is to say that some materials will show an exponential decay with the asymptote for infinite stress reaching the zero stress point (red line). Biological materials such as bone are particularly inclined to not exhibit an endurance limit as even low-stress loading can lead to microfractures in the material, which are propagated over multiple cycles [7]. Conversely, metals (especially ferrous alloys) do typically show a nonzero endurance limit as shown by Bomas et al. (Figure 1.5b) [5].

1.2.5 Viscoelastic Behavior

Outside of the typical stress–strain response of materials described earlier are a few specific cases of material behavior that warrant discussion. The first of these is the concept of viscoelasticity, or the combination of elastic and inelastic energy transfer through a material during loading. As most organic materials cannot be purely represented as an elastic solid such as the Hookean spring discussed in the previous section, viscoelastic theory helps encompass the mathematical representation of energy loss internal to the sample due to heat loss, molecular rearrangement (e.g., polymeric chain disentanglement), or compositional changes (e.g., water flux through a compressed cartilage).

Figure 1.6 shows the analogous mechanical diagram for how a material can behave as a simple Hookean spring (an elastic response), as a dampened dashpot (a viscous response), or as a serial or parallel combination of these elements (the Maxwell, Voigt, and standard linear models). The importance of these models comes into play when discussing the more complicated approximations of material behavior under load because no material can be

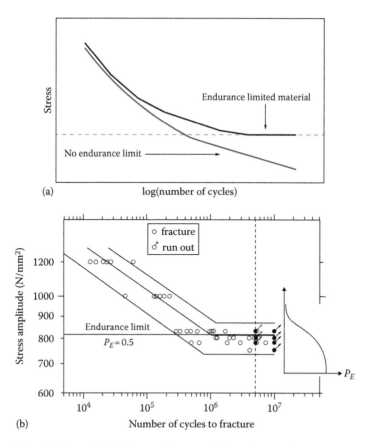

(a)

(b)

Figure 1.5 (a) Example S–N curve for materials with or without an endurance limit. (b) Graphic showing the endurance limit of bearing steel SAE 52100. (Reproduced from Bomas, H. et al., *Extremes*, 2, 149, 1999. With permission.)

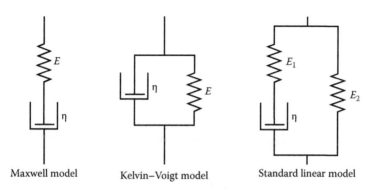

Figure 1.6 Principal constitutive models for description of viscoelasticity comprised of springs with spring constant E_n and dashpots with viscosity η.

purely elastic (some energy is lost as heat) or purely viscous (some energy will deform the specimen). In that way, the Maxwell, Voigt, and standard linear models attempt to explain how the combination of various elastic and viscous components can describe the complex behavior of a material under load. Specifically, two modes of deformation are explored by these constitutive models: creep (or the continuation of deformation after a fixed stress) and stress relaxation (or the reduction in internal stresses after a fixed deformation).

The Maxwell model shows the application of a Hookean spring with spring constant E and a Newtonian dashpot with viscosity η in series, predicting the loaded deformation of the material to show an instantaneous response and time-delayed response to any instantaneous load. Doing so describes accurately the concept of stress relaxation as there is an immediate internal stress due to an instantaneous load (spring), with a decreasing load due to the viscous recovery of the dashpot element. However, the concept of creep is poorly explained by a Maxwell model. Due to the constant force applied on the serial dashpot, there is a constant deformation associated with the viscous component. Numerically, this can be seen in the simplified differential equation for the incremental change in strain of a Maxwell solid described in Equation 1.5.

$$\frac{d\varepsilon}{dt} = \frac{\sigma}{\eta} + \frac{1}{E}\frac{d\sigma}{dt} \tag{1.5}$$

As the change in strain $(d\varepsilon/dt)$ is dependent on both the stress (σ), viscosity (η), and the change in stress $d\sigma/dt$, the instantaneous load removes the $(1/E)\,(d\sigma/dt)$ component and solves that the change in strain is equal to the value σ/η. This constant change in strain does not accurately describe how a material will deform under an instantaneous load. However, because there will be a maximum strain that can be accommodated by the material, there must also be a corresponding decay in the strain rate. To address this issue with creep in Maxwell solids, the Kelvin–Voigt model was developed.

The Kelvin–Voigt model takes the elements present in the Maxwell model and rearranges them to a parallel configuration instead of the previous serial arrangement. In this way, an instant deformation will lead to a stress on the material that gradually decreases on the dashpot component leading to a finite elastic deformation. This accurately captures the creep phenomenon; however, the Kelvin–Voigt model has an issue with the description of stress relaxation. Due to the dependence of the viscous component on the strain rate, an instantaneous deformation of a Kelvin–Voigt solid leads to an instantaneous response due to the elastic component, as well as an instant spike in the stress experienced from the viscous component. This is described by the partial differential equation seen in Equation 1.6, which shows that the stress in a Kelvin–Voigt body is proportional to the strain applied, as well as the rate of change in that strain.

$$\sigma = E\varepsilon + \eta\frac{d\varepsilon}{dt} \tag{1.6}$$

However, due to the instantaneous application of the strain, the $d\varepsilon/dt$ term is zero after the discontinuous application of the strain. This solves to the stress value in a Kelvin–Voigt model being a constant $E\varepsilon$ and not showing the characteristic stress relaxation behavior seen in fixed-strain viscoelastic materials. Contrary to the Maxwell model, which describes this phenomenon well, the Kelvin–Voigt model for the deformation of a viscoelastic body falls short of accurately modeling the relaxation of an instantaneous stress on a material.

To remedy the shortcomings of both models, the standard linear model was developed to extrapolate both the stress relaxation and the creep behavior of viscoelastic solids using the same lumped parameter model. While many interpretations exist for the standard linear model, the simplest of these is shown in the last example of Figure 1.6. In this equivalent model, the Kelvin–Voigt model has an extra Hookean spring with spring constant E_2 in series with the dashpot of the previous model. While other representations of this model can be described (e.g., the second Hookean spring is in series with the entire parallel arrangement of the Kelvin–Voigt model), the base concept is preserved: an extra elastic region is added to the Kelvin–Voigt model that can accommodate an instantaneous deformation given an instant stress and vice versa. The major implication of this simple addition is the capability of one mechanical model to explain both the creep and stress-relaxation phenomenon using the same set of partial differential equations. Using this model, Equations 1.7 and 1.8 can be derived for the standard linear model's incremental change in strain and stress, respectively. As seen using these equations, the rates of change in both the strain and stress first describe an instant response to a deformation or stress, followed either by a logarithmic growth in the deformation that asymptotes to σ/E (creep) or an exponential decay in the internal stress that asymptotes to $E\varepsilon$ (stress-relaxation).

$$\frac{d\varepsilon}{dt} = \sigma \frac{(E_1 + E_2)}{E_1 \eta} - \varepsilon \frac{E_2}{\eta} \tag{1.7}$$

$$\frac{d\sigma}{dt} = -\sigma \frac{(E_1 + E_2)}{\eta} + \varepsilon \frac{E_1 E_2}{\eta} \tag{1.8}$$

The final component of viscoelasticity needing address is the concept of time-temperature superposition (TTS), or the frequency and temperature-dependent mechanical responses of a material. The use of TTS is primarily in predicting viscoelastic behavior of materials outside of the timescales afforded by the measurement equipment discussed in Section 1.4. As these instruments can measure in the milli-Hertz to kilo-Hertz frequency regimes, viscoelastic material physics cannot be observed in the laboratory at deformation frequencies outside of this narrow range. Using the Williams–Landel–Ferry (WLF) equation, predictions can be made regarding the state of viscoelasticity in a material at an expected frequency of deformation and temperature outside of those already experimentally determined [8].

Before moving past the concept of viscoelasticity, a last measure has to be introduced: the idea of a complex modulus. Until now, the calculation of a modulus of elasticity presupposed that all energy used to stress the sample elastically deformed the specimen. However, some energy is dissipated as heat or other forms of inelastic loss within the specimen. For example, while measuring the Young's modulus, E, the measurement is in fact the combination of real (the storage modulus E') and complex (the inelastic loss modulus E'') components of a complex modulus, E^*, such that

$$E^* = E' + iE'' \tag{1.9}$$

This analogy holds true for different moduli as well, including bulk modulus (K^*), axial modulus, Lamé's first parameter (λ), Lamé's second parameter (μ)—more commonly referred to as shear modulus (G^*)—and the P-wave modulus (M).

Other types of elastic moduli such as shear modulus G (in shear a load is applied parallel to a sample's cross-sectional area) or bulk modulus K (a uniform pressure is applied across the entire sample surface area) can also be calculated depending on the type of loading for the sample. While E and G are by far the most common descriptors of elastic material deformation, many other moduli can be calculated for more exotic forms of sample loading and can be deconvolved into their respective storage and loss moduli.

1.2.6 Poisson's Ratio

Poisson's ratio, named for the nineteenth-century French mathematician, describes the relationship between the deformation in the loading axis and the deformation normal to the load axis. It can also be thought of as the coefficient of expansion on the transverse axes or the negative ratio of transverse to axial strain. Poisson's ratio varies between −1.0 and 0.5, the theoretical and practical limits for stable, isotropic, linear elastic materials. Perfectly incompressible materials have Poisson's ratios of 0.5, and include elastic rubbers. A Poisson's ratio of 0.5 is density and volume preserving. Thus, the transverse deformation is such that the density is not changed during compression. Most materials have a Poisson's ratio between 0 and 0.5 and show some signs of densification during compression or tension. Materials such as cork and Styrofoam can have Poisson's ratios of 0, meaning that there is no change in transverse directions during axial deformation. Materials with negative Poisson's ratios are called auxetic materials and actually get thicker as they are stretched. Materials such as the synthetic biomaterial Gore-Tex and some tendons [9] show auxetic properties. It is recommended that readers who want to gain a deeper understanding of the complex mechanical properties of materials, delve into more in-depth treatments of these topics and learn about tensor notation, compliance matrices, and advanced mechanics of materials from one of many great references in this area [10].

1.2.7 Isotropy

The final material property that will be discussed in this introductory chapter is the concept of isotropy in a material, or the homogeneity of the material properties as a function of loading direction. The term isotropy comes from the Greek roots *ïsos*, meaning "the same," and *tropos*, meaning "turning toward" as in aligning along a certain orientation. A material is considered to be mechanically isotropic if the mechanical response to load is uniform regardless of loading direction. However, many biomaterials (especially polymers and tissues) are loading-direction dependent, making them anisotropic. An example of this can be seen in Figure 1.7a where a typical stress–strain graph is shown for an anisotropic material, showing how the material is strongest in the longitudinal loading direction and weakest in the transverse loading direction, with corresponding decreases in mechanical robustness as the loading moves from longitudinal to transverse. In the case of fibrous materials such as plastics and tissues, the longitudinal load is typically defined as stress in the direction of the fiber orientation, and transverse load is stress perpendicular to the orientation of the fibers.

While the concept of anisotropy in materials is of great consequence in biological materials, engineered materials have usually enjoyed a reprieve from isotropy considerations, as molding (e.g., sand casting, injection molding) for these materials typically introduces uniform properties regardless of direction. However, with the advent of rapid prototyping and 3D printing, isotropic behavior in materials is quickly becoming a critical issue in design and manufacturing of biomedical devices. An example anisotropic biomaterial is the fused filament fabrication (FFF) of poly(lactic acid) specimens [2]. In this study, Shaffer et al. show how they can reduce anisotropy in poly-lactic acid (PLA) dog bones by using gamma radiation to induce covalent linkages between polymer chains. The method involves the printing of dog bones with variable degree of fiber orientation (Figure 1.7b) and studying their as-printed mechanical properties (Figure 1.7c) and the same properties after gamma irradiation (Figure 1.7d). While this demonstrates only one way of reducing anisotropic behavior of materials that are manufactured with direction-dependent mechanical properties, reducing anisotropy in biomaterials is of great interest in applications where load directionality is not guaranteed.

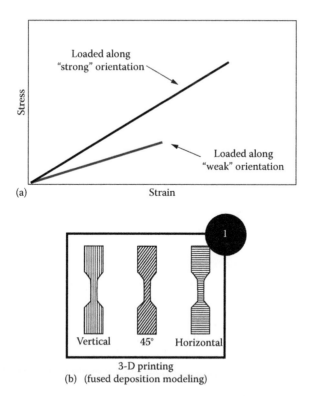

Figure 1.7 (a) Representative loading-direction dependency of anisotropic material properties. (b) Fused filament fabrication (FFF) of varying degree-of-fiber-orientation PLA dog bones and their corresponding stress–strain characteristics. *(Continued)*

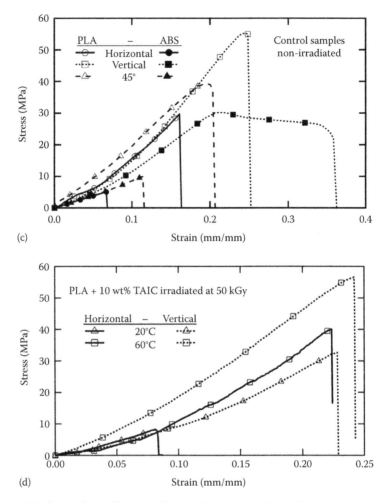

Figure 1.7 (Continued) (c) Before and (d) after 50 kGy of gamma irradiation. (Reproduced from Shaffer, S. et al., *Polymer*, 55, 5969, 2014. With permission.)

1.3 COMMON ENGINEERING BIOMATERIALS

Tailor-made biomaterials are ones that have been engineered for specific purposes or have traits that make them appropriate for a given application. While there are a great number of these materials utilized in biomedical applications, we will present a few key metals (stainless steel, titanium and its alloys, and cobalt chromium alloys), ceramics (alumina, zirconia, and hydroxyapatite), and polymers (PEEK, PE, and PLGA).

Perhaps the oldest of the modern medical alloys, stainless steel has been used across a number of implantable and surgical devices. Composed primarily of iron, chromium, and nickel, stainless steel has been shown to be biocompatible, corrosion-resistant, and has the mechanical properties needed for a range of applications. The most common type of stainless steel for bio-applications is 316 stainless (generally, 18% chromium, 14% nickel, and 2.5% molybdenum) followed by 304 stainless (generally, 19% chromium and 10% nickel). There have been at least 325 devices made from 316 stainless steel compared to 82 made from 304 stainless steel [11].

Seeking to improve upon the corrosion resistance and biocompatibility of the stainless steel alloys, commercially pure titanium was first used in a medical device in 1940 [12]. The greater conformity of titania (titanium oxide) to the bulk titanium over chromium oxide to stainless steel and the lack of nickel account for both improvements in biocompatibility and corrosion resistance. Commercially pure titanium has been used in at least 321 medical devices [11] in a number of different kinds of devices and applications. However, the mechanical properties of commercially pure titanium, as seen in Table 1.1, led to the development of titanium alloys to create materials with the biocompatibility/corrosion resistance of commercially pure titanium but with improved stiffness. The most common of the titanium alloys is Ti–6Al–4V, which as its name suggests contains 6% aluminum and 4% vanadium. This alloy is used in at least 1463 medical devices, demonstrating its wide appeal as a biomaterial [11]. The final titanium-based biomaterial of note is Nitinol, a near equiatomic alloy of titanium and nickel. Nitinol was developed in the 1960s [13] and while it exhibited unique shape memory properties, it was not until the late 1970s that it found a role as a biomaterial [14]. The shape memory properties of nitinol allow stents to self-expand instead of being balloon expanded (which brings a risk of blockage and stroke to the patient), allows for easier introduction through a catheter, and offers a nonoperative approach to surgeries [15,16]. While nitinol has been used in at least 547 medical devices [11], the addition of that much nickel to the titanium has led to concerns regarding corrosion resistance and nickel leaching of the alloy [17].

Table 1.1: Overview of Mechanical Properties of Engineered Biomaterials

	Modulus (GPa)	Yield Strength (MPa)	Tensile Strength (MPa)	Elongation at Break (%)	Density (g/cm³)
316LVM[a] (annealed)	188	190	490–800	45	8
304[a] (annealed)	193	330	590	64	8
CP titanium[a] (annealed)	102	340	430	28	4.5
Ti–6Al–4V[a] (annealed)	113	880	950	14	4.4
Nitinol[a] (above A_f)	75	560	754–960	15.5	6.4
MP35N[a] (annealed)	232	414	931	70	8.4
F1537[a] (annealed)	241	585	1035	25	8.3
Alumina[b]	360–400	—	170–250	—	4
Zirconia[b]	170–260	—	124–1200	—	5.5
HA[b]	73–120	—	—	—	3
PEEK[b] (unfilled)	2.9–5	—	91–111	14–38	1.1
PEEK[b] (carbonfiber)	14–25	—	95–226	1–3.24	1.4
Polyethylene[b] (low density)	0.24–0.358	—	11–21	335–650	0.8
Polyethylene[b] (cross-linked)	1–1.3	—	36–62	10–440	—
PLGA[b]	1.2–2.9	—	41–55	2.3–9.5	1.3

[a] Matweb: Material Property Data [26].
[b] From Reference 11.

Beyond stainless steels and titanium/titanium-based alloys, cobalt chromium alloys have also been used in biomedical application for their high corrosion resistance and high elastic modulus. The high amounts of chromium in the alloy account for its "good" corrosion resistance, while cobalt provides the high elastic modulus. For cardiovascular applications, MP35N (ASTM F562, cobalt with 35% nickel, 20% chromium, and 10% molybdenum) is quite popular, whereas for orthopedic applications, a cobalt chromium molybdenum alloy (ASTM 1537, cobalt with 26% chromium and 5% molybdenum) is quite popular. In terms of usage in medical devices, at least 45 devices use MP35N, while at least 145 devices use the ASTM F1537 alloy [11]. Despite the high corrosion resistance of these alloys, there have been recent concerns with the biocompatibility of cobalt chromium wear debris in orthopedic applications as well as concerns with crevice corrosion.

The issues of biocompatibility of wear debris as well as corrosion concerns have made ceramics more attractive as an alternative material to metal alloys in orthopedic applications, despite the brittleness inherent to ceramics. The covalent bonds of ceramics make them highly resistant to in vivo corrosion; ceramics require highly aggressive environments for corrosion to occur [18], and the lack of unbonded metal keeps the wear particles from being a biocompatibility risk [19]. Due to the high modulus, low wear, and brittleness of ceramics, most of their applications are regulated to orthopedics (e.g., joint replacements) or for device bone interfaces in the case of hydroxyapatite.

Alumina (aluminum oxide) was first used in a conceptual medical device in the 1930s but it was not until the 1960s when material quality and processing made it feasible to be used in an orthopedic application [20]. Since then, the number of medical devices utilizing alumina have increased to 93, the vast majority of them being the femoral head component of a hip replacement system [11]. As the field of ceramic processing improved, zirconia (zirconium oxide), which has better strength and toughness than alumina, became a viable material for orthopedic devices in the late 1970s. Since then, at least 79 devices utilizing zirconia have been reported; like alumina, a majority of these devices have been femoral heads for hip replacement devices [11]. While the high strength and hardness of alumina and zirconia have made them attractive to be used in orthopedic applications, their chemical stability precludes them being used as a degradable interface. Hydroxyapatite ($Ca_5(PO_4)_3(OH)$) is a degradable ceramic that first saw use as a biomaterial in the 1970s and since then has been used in 237 medical devices, predominantly as a degradable coating [11].

Polymers, like ceramics, are composed of covalently bonded atoms and the lack of metals or heavier elements means that corrosion and metal leaching is not typically a concern when they are used as a biomaterial. While the wide range in modulus and mechanical properties coupled with the general biocompatibility of polymers make them attractive for a number of different applications, polymers do carry a risk of degradation and leaching residuals from the polymerization process [21]. While there are a number of different polymers used for biomedical applications, the three that cover the spectrum of uses are poly-ether-ether-ketone (PEEK), polyethylene (PE), and poly(lactic-co-glycolic acid) (PLGA).

PEEK was shown to be biocompatible in the late 1980s [22] and since then has been used in a number of applications spanning orthopedic and trauma applications as the material can be processed to match cortical bone [23]. While the most popular application for PEEK is in spinal implants, PEEK has been used in at least 779 medical products that also include anchors, screws, and plates [11]. While PEEK is a rigid polymer with properties very similar to bone, polyethylene generally has lower stiffness, making it more suitable for applications where a softer material is desired. As a result, polyethylene has been used in at least 1043 devices [11] across a broad range of medical device areas. It is worth noting that with the advent of radiation cross-linking and vitamin E fortification, polyethylene has been used in orthopedic applications as a bearing surface material [24]. PLGA has been proposed as a degradable biomedical material as early as the 1990s [25], low modulus and hydrolytic degradation making it substantially different from PEEK and polyethylene. This has led to PLGA being used in at least 44 medical devices [11] for applications ranging from screws, pins, and plates.

This has been a brief summary of engineered biomaterials and has only covered a small fraction of the engineered biomaterials that are available. The greatest strength of engineered biomaterials is that materials are always improving and new materials can be developed to address specific needs/applications.

1.4 CHARACTERIZATION OF MECHANICAL PROPERTIES

The characterization of the mechanical properties of biomaterials is such an open-ended statement that we cannot begin to do the topic justice with any reasonable depth in the limited space we have here. Therefore, we refer the reader to a host of books and articles that delve into deeper biomechanics in many narrow areas including the studies of various biomaterials described in Section 1.3 with detailed focus on bone and cartilage as case studies. We also delve into broad categories of testing ceramic, metal, and polymer synthetic biomaterials. Much of the focus herein revolves around bulk materials properties, cognizant of the fact that most biological materials exhibit highly anisotropic mechanical properties that differ dramatically across time and length scales. This section breaks down mechanical characterization relative to the tool used and attempts to paint a broad overview of the power of structure–property relationships that can be gleaned with macroscopic and microscopic mechanical testing.

1.4.1 Universal Testing Machine (UTM)

The stress–strain curve, an example of which is shown in Figures 1.2 and 1.3, is the common resulting data provided by the UTM, which is also known as a tensile tester or a materials test frame, or a load frame. Many different companies make load frames and UTM components with some variation on the following underlying science: a material of known geometry is affixed into a test structure such that it can be deformed in a known amount, in a known direction, by a known load, at a known deformation rate, at a known temperature, and at a known humidity.

Many variations on this basic theme exist and will be touched upon here. The known amount of deformation (strain) can be accomplished in tension, compression, torsion, shear, or complex combinations thereof such as three-point bending, four-point bending, biaxial tension, extension and torsion, or hydrostatic compression. In this text we will focus on uniaxial loading in tension and compression and delve into more complex deformation modes in Section 1.4.2 which are seen in dynamic mechanical analysis (DMA), where we move from quasistatic deformation to dynamic oscillatory deformation.

A UTM consists of several components. The load frame is typically constructed from strong metal or composite support structures. Attached onto these support structures are typically large metals screws or (servo-)hydraulic encasings that enable the (usually) vertical deformation. Detailed information about different tensile testers can be found in various places online, from manufacturer literature to print [27]. A schematic of a screw-driven tensile tester with tensile clamps (grips) is shown in Figure 1.8. A UTM also contains the crosshead, which moves vertically and at constant rates to track how far the sample has moved

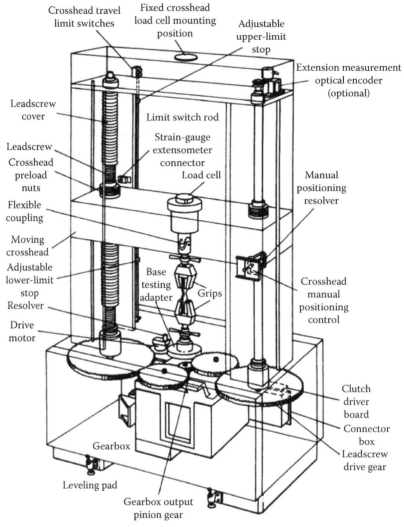

Figure 1.8 Typical schematic of a UTM. (Reproduced from Davis, J.R., *Tensile Testing*, 2nd edn., ASM International, Materials Park, OH, 2004. With permission.)

since the start of the test. Many systems utilize secondary means of tabulating deformation through video extensometers, laser extensometers, or simple strain gauges to obtain more precise measurements. Load and displacement (measured directly by the machine) are converted into stress and strain (given the input test environment and sample geometries) and recorded for postprocessing and further study. Due to specimen variability, it is generally recommended to test multiple specimens and average the results before reporting a value for any given material property; a great resource for the determination of sample size in a random sample is ASTM standard ASTM-E122 [28]. These tests are conducted under the same loading conditions and used to compute averages for yield stress, ultimate tensile stress, strain-to-failure, and any other computed parameters such as Young's modulus or toughness.

1.4.1.1 Compression

When a uniaxial compressive load is applied onto a flat, bulk sample of known geometry, it exerts a measureable force on the material that is detectable by a load cell (also known as a force transducer) that is situated normal to the sample. What is actually measured is an electronic signal, which can be a change in voltage, current, or frequency, depending on the type of load cell and circuitry used. The electronic signal will have been calibrated during instrument setup to correlate to specific loads. The force is computed into a stress by dividing by the cross-sectional area of the material that is being deformed. The strain that is calculated from the initial height and deformed heights between the compression platens using the initial cross-sectional area is engineering strain.

Herein lies the first of many problems in obtaining accurate measurements from biological materials. If the sample loaded between compression platens is not preprocessed (cut, sectioned, microtomed, etc.) to have parallel top and bottom faces, the load applied to the sample is no longer strictly uniaxial. Often, preparing samples into the necessary geometry destroys valuable structural properties of the material. If the biological material is anisotropic (and it almost certainly will be at some length scale), the direction of loading of the material becomes relevant. The total sampling plan increases dramatically if biomaterials must be harvested and tested at different angles relative to known tissue anisotropies, for example. If the material is heterogeneous, it becomes difficult to ascertain whether the load is actually applied uniformly uniaxially across a cross section of known material or across some nonplanar region within the material that is not perpendicular to the loading axis. Moreover, if the material has internal structures, such as in biological foams, differences in cross-sectional area perpendicular to the axis of deformation play an enormous role in the shape of the resulting stress–strain curve. If the material is not well adhered to one or both of the compression platens, slipping can occur and disrupt proper data collection. If adhesives are used to attach materials to the compression platens, they have the ability to alter the mechanical properties by creating material/ adhesive composites that exhibit different properties from bulk materials. Additionally, under a compressive load, the weakest part of the cross section of the material will deform first. At the onset of a test, this is usually on one of the surfaces that are not uniformly flat. Thus, a relatively small preload (e.g., 0.1 N for samples tested to 10 N) can be applied to help overcome surface imperfections and misalignments. Assuming that each of these obstacles can be avoided to properly conduct a test, only then is a compressive stress–strain graph a useful quantitative representation of the specimen's material properties. From these compressive stress–strain data, the user is then able to estimate the Young's modulus, yield stress, ultimate compressive strength, toughness, failure mechanics (e.g., brittle vs. ductile failure), and countless other parameters using the techniques discussed in Section 1.2.

1.4.1.2 Tension

When a uniaxial tensile load is applied onto a bulk sample of known geometry, the machine exerts a measurable force onto the sample that is detectable by the load cell, again typically situated above the sample. At the beginning of the quasi-static test, the UTM is sometimes able to approximate the Young's modulus as the sample is deformed in the linear elastic regime. Any deformation in this regime should have little to no hysteresis and a linear slope until the yield stress of the material. Materials of different stiffness exhibit different properties as a function of temperature, humidity, and frequency of deformation. It is critical to track these parameters and report them during tests. UTM tests should be run at specific temperatures because the properties of many materials, especially materials highly optimized to serve as mechanical structural components in the body, change dramatically as a function of temperature. Some UTM fixtures allow for fluid baths and recirculators that enable testing at 37°C in biological fluids. The problems that emerge with these setups include longer setup times and corrosion issues that damage testing equipment if long hydrated tests are conducted, especially in saline environments. In these cases of testing within a fluid bath, the properties of the fluid, such as the concentration of a phosphate-buffered saline (PBS) and the pH of mixture are relevant. The most common shape for synthetic biomaterials tested in tension is the ASTM Type IV or Type V dog bone sample, which was designed to minimize stress concentrators and localize most of the deformation away from the grip regions and onto the gauge section. While this works well for low strain capacity materials, higher strain materials begin to deform in the grip section as well, which can limit the accuracy of the crosshead in determining the strain. In this case, it is imperative to have an additional setup to accurately measure strain. In the cases of testing biological materials such as tendons, the standards should be used as starting points, but are likely not perfectly relevant for all possible setups. It is important to understand the intent behind the standards and attempt to reflect that in the test setup that is designed (e.g., ASTM D638 for tensile testing of plastics) [29]. From these tensile stress–strain data, the user is again able to estimate the Young's modulus, yield stress, ultimate tensile strength, toughness, failure mechanics (e.g., brittle vs. ductile failure), and other parameters using the techniques discussed in Section 1.2.

1.4.2 Dynamic Mechanical Analysis

Dynamic Mechanical Analysis (DMA) is a technique used to characterize the temperature and frequency dependence of materials. In DMA, an oscillating force is applied to a material at a known frequency in a known sample geometry and the resulting strain stress and strain are calculated. Since the material is kept in the linear elastic regime, the stress–strain relationship is linear, yielding the modulus of interest (typically E^* or G^*). A load cell, which can range from 1 mN to 40 N, is used to apply the appropriate engineering stress in stress-controlled modes. In strain-controlled modes, samples are deformed to a known deformation (0.2% for example) and the stress is calculated. Strain can be measured in many ways but is often tabulated via a linear variable transducer, or a moving magnet in a magnetic field that can indicate deformation with nanometer-scale precision. Modes of deformation in a DMA can include tension, compression, three-point bend, four-point bend, single cantilever, dual cantilever, shear, liquid shear, and perhaps others. We will describe tension, compression, and shear modes in this section and leave it to the reader to find other literature describing other deformation modes as needed. During a temperature sweep, the DMA repeatedly strains the materials at a known frequency, or frequencies in multiplexed modes, as the thermal chamber surrounding the sample varies in temperature (from ranges as wide as −80°C to 500°C depending on the test). For biological materials, a thermal sweep contained within the liquid water thermal region (0°C–100°C) is often most relevant at a range of frequencies. Often, when a sample is cooled it should be reclamped at the lowest test temperature, re-equilibrated to that temperature, and then tested to ensure good fixation that would account for thermal contraction. For high strain materials, or materials that undergo dramatic changes in properties over the test temperature range, often it is necessary to glue specimens in place with an adhesive that is subsequently removable with moderately aggressive solvents. A certain force must be applied to deform the sample a known amount and the measurement of this strain often lags (in phase) behind the applied stress. For perfectly elastic solids that follow Hookean behavior, these are perfectly in phase. For a purely viscous fluid, the phase lag, δ, will be 90° of strain with respect to stress. When the phase lag is 0°, the stress as a function of time is proportionate to the strain:

$$\sigma(t) = E\varepsilon(t) \quad \text{or} \quad \sigma_0 \sin(\omega t + \delta) = E\varepsilon_0 \sin \omega t \quad \text{when } \delta = 0 \qquad (1.10)$$

Similarly, when the phase lag (δ) = 90°:

$$\sigma(t) = \eta \frac{d\varepsilon}{dt} \quad \text{or} \quad \sigma_0 \sin(\omega t + \delta) = E\varepsilon_0 \cos \omega t \quad \text{when } \delta = \frac{\pi}{2} \qquad (1.11)$$

where
 $\sigma(t)$ is the time-dependent stress
 E is the elastic modulus
 $\varepsilon(t)$ is the time-dependent strain
 ω is the frequency of strain oscillation
 δ is the phase lag
 η is the viscosity

For viscoelastic materials, such as materials near a glass transition temperature and many biological materials, the stress as a function of time can be represented by a combination of elastic springs (Equation 1.10) and viscous dashpots (Equation 1.11) in series or parallel as described in different arrangements in Section 1.2 with the Maxwell, Kelvin–Voigt, and standard linear models. As we described in Equation 1.9, the elastic modulus contains both real and complex portions. The real portion is E' and the complex portion is E'':

$$E' = \frac{\sigma_0}{\varepsilon_0}\cos\delta \quad E'' = \frac{\sigma_0}{\varepsilon_0}\sin\delta \qquad (1.12)$$

The phase lag then simplifies to the ratio of the loss modulus (E'') to the storage modulus (E'):

$$\delta = \arctan\frac{E''}{E'} \qquad (1.13)$$

The DMA then outputs a curve that measures the time-varying moduli and the phase lag as a function of temperature and frequency. One of the most useful applications of DMA is the application of a multiple-frequency sinusoidal deformation, with a superposition of different waves that vary in frequency across a single decade (such as 1, 2, 5, and 10 Hz). Those oscillations can be decoupled into their respective modes such that a user could obtain information about materials properties at multiple frequencies during a single temperature sweep. This sweep is especially useful when studying complex synthetic biomaterials that may contain combinations of glass transitions, melt transitions, and dissociation transitions that may even be overlapping. The WLF equations briefly described in Section 1.2 show how materials properties change as a function of frequency. Time—temperature superposition curves can be made that indicate the location of thermal transitions and the modulus on either side of those transitions. In some cases, polymers exhibit a shift in properties (such as a shift in the glass transition temperature of 6.5°C per decade of frequency) at the same temperature. An example of this shift would manifest in a material that is being perturbed by an oscillatory 1 Hz wave at 37°C and is suddenly exposed to an oscillatory 1 kHz wave at 37°C; this perturbation means that the material would have an effective increase in glass transition temperature of nearly 20°C. Moreover, if materials have been optimized for physiological conditions and subsequently experience dynamic loads at unexpected frequencies, the materials could move to ranges where they are much weaker and less effective. For instance if tendons, cartilage, and other biological materials are deformed outside of their desired parameter spaces, they can lose some of their great mechanical properties; by using DMA, some of this information can be extracted and analyzed before load-bearing materials are implanted.

1.4.2.1 Setting up a DMA Test

In tension, samples are cut into strips and clamped into the DMA system. Great care must be taken to affix the materials into the clamps. Beyond using adhesives, other techniques can be employed to wrap grip regions with double-sided sandpaper that aid in fixation. Before the oscillatory tensile test begins, it is important to add a static preload to the material, so that the materials never enter into a buckling state, wherein one surface experiences tensile forces and the other compressive forces. This will limit the DMA's ability to accurately report stresses at known strains. Once the static preload is set, a dynamic oscillatory strain is placed on the material such that the material is always in tension. This is perhaps one of the most critical steps that beginner users may not always do correctly, which can lead to erroneous data.

In compression, preload is used in much the same way as it is in tension, to ensure that the sample always remains under some form of compression. The preload exerts a known static load on the material and proceeds to dynamically modulate stress around this preload. If the platens ever lose contact with the material, the instrument can no longer measure the material's response to load and the test is not relevant. Critically, however, materials behave differently under different preloads. Take a biological foam, for example: if the preload sufficiently alters the pore structure, the time-varying dynamic oscillatory stress–strain response could be very different than from minimally preloaded samples.

In shear, preload is not relevant on the axis of deformation, but rather perpendicular to the axis of deformation. Shear samples are typically clamped vertically with a known force. The force with which the samples are clamped, relative to their density, geometry, size, and modulus, will affect how the material shears through its cross section.

Thus in DMA literature, what is stressed is the importance to measure constant density homogenous materials. This literature accurately describes few biological materials and only some synthetic biomaterials. Therefore, when testing complex biological materials, it is critical to understand the underlying mechanics of the test environment and design a test protocol with the appropriate sample size to suss out the desired information. It serves no purpose to put a sample in the chamber, push a few buttons, and report a curve of modulus as a function of temperature or frequency if the underlying test is invalid. Using DMA to study dynamic mechanics of biomaterials is an exciting area that is fraught with uncertainty in achieving useful relevant data but can be very rewarding for the design of next-generation solutions of grand challenges in healthcare, medical devices, and biomedicine.

1.4.3 Other Mechanical Analyses of Biomaterials

Other techniques such as atomic force microscopy (AFM), nanoindentation, rheometry, impact testing, and others can be used to detect specific mechanical properties in what is an infinite parameter space of composite structures, time, static preload forces, dynamic forces, frequencies of deformation, and environmental conditions such as temperature, humidity, or atmospheric pressure. However, it is important to remember that ex vivo mechanical tests, whether on biological samples or synthetic biomaterials, can aid in building a representative model for devices or tissue behavior in or around the body. These models are limited by their inputs and can never truly simulate what happens in situ, especially with incorrectly acquired experimental data. Thus, mechanical tests should be carefully designed to study the design of new materials or to enable predictive models about the behavior of biological materials. Data from conducted tests of biological samples, unlike from precision-machined homogenous synthetic materials, should not be considered generalizable "facts" for other biomaterials, but rather anecdotes from a place and time and inspiration for the continued development of biomaterials. In this light, mechanical testing of biomaterials is a critical and logical tool for the continued understanding and development of biological materials, bio-inspired synthetic materials, and purely synthetic biomaterials.

DISCLAIMER

The mention of commercial products, their sources, or their use in connection with material reported herein is not to be construed as either an actual or implied endorsement of such products by the Department of Health and Human Services. The opinions and/or conclusion expressed are solely those of the authors and in no way imply a policy or position of the Food and Drug Administration.

REFERENCES

1. C.M. Yakacki, S. Willis, C. Luders, K. Gall. *Advanced Engineering Materials*, 2008, *10*, 112.

2. S. Shaffer, K. Yang, J. Vargas, M.A. Di Prima, W. Voit. *Polymer*, 2014, *55*, 5969.

3. B.Z. Jang, D.R. Uhlmann, J.B.V. Sande. *Journal of Applied Polymer Science*, 1984, *29*, 3409.

4. H. Peterlik, P. Roschger, K. Klaushofer, P. Fratzl. *Nature Materials*, 2006, *5*, 52.

5. H. Bomas, P. Mayr, T. Linkewitz. *Extremes*, 1999, *2*, 149.

6. V. Levitin. *Interatomic Bonding in Solids: Fundamentals, Simulation, Applications*, 2013, John Wiley & Sons, Weinheim, Germany.

7. D.R. Carter, W.E. Caler, D.M. Spengler, V.H. Frankel. *Acta Orthopaedica*, 1981, *52*, 481.

8. R.M. Christensen. *Theory of Viscoelasticity: An Introduction*, 1971, Academic Press, New York.

9. R. Gatt, M. Vella Wood, A. Gatt, F. Zarb, C. Formosa, K.M. Azzopardi, A. Casha et al. *Acta Biomaterialia*, 2015, *24*, 201.

10. A.P. Boresi, R.J. Schmidt, O.M. Sidebottom, *Advanced Mechanics of Materials*, 1993, John Wiley & Sons, New York.

11. ASM International. *ASM Medical Materials Database*, 2016.

12. R. Bothe, K. Beaton, H. Davenport. *Surgery, Gynecology and Obstetrics*, 1940, *71*, 598.

13. W.J. Buehler, F.E. Wang. *Ocean Engineering*, 1968, *1*, 105.

14. M. Simon, R. Kaplow, E. Salzman, D. Freiman. *Radiology*, 1977, *125*, 89.

15. C.T. Dotter, R. Buschmann, M.K. McKinney, J. Rösch. *Radiology*, 1983, *147*, 259.

16. A. Cragg, G. Lund, J. Rysavy, F. Castaneda, W. Castaneda-Zuniga, K. Amplatz. *Radiology*, 1983, *147*, 261.

17. U.S. Food and Drug Administration. *Food and Drug Administration, Center for Devices and Radiological Health,* January 2005, *13*.

18. C.V. Nakaishi, L.K. Carpenter. Ceramic corrosion/erosion project description (No. DOE/METC/SP-111). Department of Energy, Morgantown Energy Technology Center, Morgantown, WV. 1981.

19. K. Kato, H. Aoki, T. Tabata, M. Ogiso. *Artificial Cells, Blood Substitutes and Biotechnology*, 1979, *7*, 291.

20. W. Rieger. *World Tribology Forum in Arthroplasty*, 2001, pp. 283–294.

21. E. ISO. German Version: DIN EN ISO, 2008, 10993.

22. D. Williams, A. McNamara, R. Turner. *Journal of Materials Science Letters*, 1987, *6*, 188.

23. S.M. Kurtz, J.N. Devine. *Biomaterials*, 2007, *28*, 4845.

24. E. Oral, S.D. Christensen, A.S. Malhi, K.K. Wannomae, O.K. Muratoglu. *The Journal of Arthroplasty*, 2006, *21*, 580.

25. L. Cima, D. Ingber, J. Vacanti, R. Langer. *Biotechnology and Bioengineering*, 1991, *38*, 145.

26. Matweb: Material Property Data. 2016.

27. J.R. Davis. *Tensile Testing*, 2nd edn., 2004, ASM International, Materials Park, OH.

28. ASTM E122-09. *Standard Practice for Calculating Sample Size to Estimate, with Specified Precision, the Average for a Characteristic of a Lot or Process*, 2011 ASTM International, West Conshohocken, PA.

29. ASTM D638-14. *Standard Test Method for Tensile Properties of Plastics*, 2014, ASTM International, West Conshohocken, PA.

2 Nonlinear Mechanics of Soft Biological Materials

John F. Eberth and Tarek Shazly

CONTENTS

2.1 INTRODUCTION: CHARACTERISTICS OF SOFT BIOLOGICAL MATERIALS

Any substance or mixture that has been produced on the surface of or within a living organism is considered a biological material. In contrast, biomaterials constitute a broader classification and include any natural or man-made materials that interact with living systems (Park and Lakes, 2007). To retain maximal relevance to disease and medicine, we will concentrate here on those materials produced by the multicellular eukaryotic organisms of the animal kingdom. These materials provide structure to the organism, constitute essential organs, and enable the living organisms to interact with the environment through both sensory and stimulatory actions. It is not surprising that formal studies of the mechanics of soft biological materials have had a long history. For example, in Aristotle's *The Parts of Animals*, he suggests the causal relationship between the constituents of animal tissues and mechanical function. For example, when comparing the esophagus and windpipe of man, he observes that

> This esophagus is of a flesh-like character, and yet admits of extension like a sinew. This latter property is given to it, that it may stretch when food is introduced; while the flesh-like character is intended to make it soft and yielding, and to prevent it from being rasped by particles as they pass downwards, and so suffering damage. On the other hand, the windpipe and the so-called larynx are constructed out of a cartilaginous substance. For they have to serve not only for respiration, but also for vocal purposes; and an instrument that is to produce sounds must necessarily be not only smooth but firm.

The Parts of Animals—**Book II, Part 3**

Based on archeological evidence, informal studies of biological materials and their mechanical properties likely predate recorded history. For example, rawhide and tanned leather have both been used since Paleolithic times for protection and clothing. The primary difference between these two, and their application, is rooted in alterations in mechanical properties that are achieved through the tanning process that permanently alters the protein structure of the animal skin (Snodgrass, 2015).

The mechanics of soft biological materials can be observed at many levels of size and scale (e.g., atomic, molecular, cellular, tissue, organ, or organism) and each provides unique information about the state of health or disease of that organism. In an effort to reduce the complexity of mathematical models of tissue mechanics, a bound can be applied to the lower scale of size and the biological materials discussed as a continuum (Fung, 1993). As such, the mechanical behavior of soft biological materials can be modeled as a continuous mass and represented by a continuous function in space. In this way, very large systems of molecules and atoms can be modeled and described by a handful of equations and the essential problems in mechanics reduced to a few postulates (Humphrey and Delange, 2004). Accordingly, every problem in continuum mechanics involves equations that pertain to the applied loads (e.g., stresses, tractions, forces), kinematics (i.e., strain, stretch), equilibrium/balance relations (i.e., momentum, mass, energy, entropy), constitutive relations, and boundary/initial conditions (Humphrey, 2002).

Biological materials are largely composite structures consisting of both cells and the extracellular matrix (ECM) components. Collectively these constituents comprise the tissues and organs of the human body and their relative abundance, and interactions within the material influence the physical properties. Of these physical properties, we are most interested in are those that result from the application of forces or displacement and pertain to the calculation of stresses and strains. These properties include measures of tensile and compressive stiffness, strength, and energy storage. In brief, stress is defined as the force acting over an oriented area and cannot be measured directly but rather must be calculated. Strain is a measure of the deformation of the material that is normalized to a reference dimension. Both stresses and strains, and their many permutations, are central to the field of continuum mechanics and thus essential concepts in the current chapter. It is worth noting here that careful consideration and reporting of the exact measure of stress or strain (Tables 2.1 and 2.2) is important to disseminating knowledge gleaned from experiments.

Table 2.1: Definitions of Stress

Name and Symbol	Alternate or Related Names	Defined In	Utility in Soft Tissue Mechanics
Cauchy stress σ	Stress tensor, true stress	Area-deformed, force-deformed configurations	Interpretation of experiments, in vivo stress
First Piola–Kirchhoff stress $\mathbf{P}$	Nominal stress, engineering stress (one-dimensional)	Area-reference, force-deformed configurations	Experimentally convenient, useful for small strains
Second Piola–Kirchhoff stress $\mathbf{S}$	Material stress	Area-reference, force-reference configurations	Constitutive formulations, fictitious

Table 2.2: Definitions of Strain

Name	Alternate or Related Names	Defined In	Utility in Soft Tissue Mechanics
Green strain **E**	Green–Lagrange strain, Lagrangian strain, St. Venant strain	Reference configuration	Measure deformation part of motion, symmetric
Linearized Green strain ε	Engineering strain (one-dimensional)	Reference configuration	Limited utility, small strains only, symmetric
Deformation gradient **F**	Stretch ratio λ (principal only)	Reference and deformed configurations	Measures rigid body rotations and strains
Right Cauchy–Green strain **C**	Green's deformation tensor	Reference configuration	Independent of rigid body motion, symmetric
Left Cauchy–Green strain **B**	Finger's deformation tensor	Deformed configuration	Independent of rigid body motion, symmetric

Soft tissues are highly deformable when a mechanical load is applied. In fact, the differences in configuration between the reference (prior to loading) and deformed (as a result of loading) states can be large. Large deformations are best described within the framework of the finite strain theory of continuum mechanics where a clear distinction exists between these two configurations. Examples of soft tissues of the body include tendons, ligaments, skin, muscles, nerves, membranes, and blood vessels. In contrast, hard tissues, such as bone and dentin, contain a highly mineralized ECM that limits deformation. Those hard tissues are well studied within the context of infinitesimal or small strain theory.

Most tissues are regularly bathed in bodily fluids to maintain hydration and facilitate nutrient exchange. The bodily fluids have an influence on both the acute and chronic mechanical behavior of these tissues. For example, water bound within the ECM results in soft tissue that is practically incompressible. Unbound water exchanged with the surroundings contributes to a viscous mechanical dampening and can affect the overall stress–strain behavior. Moreover, living tissues are dynamic in form and function and the constitution can change over time. For example, synthetically active cells such as dermal fibroblasts can generate functional proteins that contribute to the tissue's mechanical structure while enzymes such as the matrix metalloproteinases can act on the order of minutes to degrade the ECM. The continuum theory of "growth and remodeling" can be used to describe these dynamic structural changes. In the current chapter, however, a series of characteristics common to soft tissues undergoing large deformations are described with only time-invariant characteristics (i.e., no creep, stress relaxation, or growth and remodeling). Accordingly, soft biological materials generally take on the following characteristics:

■ *Nonlinearity*: Nonlinearity refers to the nature of the relationship between stress and strain for soft biological materials (Figure 2.1a). This behavior is not proportional and therefore cannot be characterized by a linear modulus. The unwinding of composite long-chain polymeric structures during stretching of the key structural proteins results primarily in an entropic response. The stress–strain response appears more compliant at low strains and stiffer at high strains. Additional contractile stresses can be generated through cells that contain activated stress fibers.

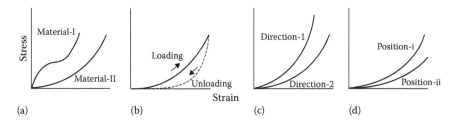

Figure 2.1 Stress–strain characteristics of soft biological materials. (a) An illustration of the nonlinearity of two different materials, I and II. Material I behaves like an elastomer and Material II like a general long-chain polymer. (b) Hysteresis is demonstrated by the loading and unloading curves that are not coincident. (c) Anisotropy of a material being pulled in two different directions, 1 and 2. (d) A tissue is heterogeneous when the mechanical response depends on the position i or ii of mechanical testing.

- *Pseudoelasticity*: A purely elastic material does not dissipate energy upon deformation, and therefore the loading and unloading curves coincide. Due to the viscous nature of the ground substance within the ECM and the movement and alignment of proteins within the bulk of the material, hysteresis exists between the loading and unloading curves and therefore indicates a loss of energy (Figure 2.1b). The interpretation is further complicated in ex vivo testing where the number of cycles of loading, the maximum loading encountered on the previous cycle (Mullins effect), and the rate of strain can bias the stress–strain curve. Collectively, these behaviors that demonstrate time, loading/unloading, or cycle-dependent strain can be considered as viscoelastic, that is, having the properties of both viscous and elastic materials. Accounting for viscoelasticity, however, adds significant complexity to mathematical formulations of soft tissue mechanics. Instead, the behavior is described and coined by YC Fung within the framework of pseudoelasticity by studying, either the loading or unloading curves of the tissue, and by following careful and well-designed experimentation that includes multiple cycles of preconditioning (Fung, 1993).

- *Anisotropy*: Although certain soft biological materials may exhibit mechanical behavior that is independent of the direction of loading (isotropic), most composite tissues have mechanical behavior that is directionally dependent (anisotropic). This directional dependency happens as a result of the fibrous nature of many soft biological materials having fibers aligned in the load-bearing direction to optimize in vivo functionality. The mechanical properties can also be described as distinct in one (transverse isotropy) or more (orthotropic) orthogonal directions (Figure 2.1c).

- *Heterogeneity*: Regional differences in material properties and composition are commonly demonstrated by soft biological tissues even within the length scales implicit in the continuum mechanics approach. Any tissue that has mechanical properties that depend on the region being tested is considered heterogeneous. These region-specific properties are endowed by compositional variations within tissues (Figure 2.1d).

Simple inflation and extension of a blood vessel can be used to illustrate the mentioned mechanical behaviors common to soft tissues. In a representative case (Prim et al., 2016), we observed and recorded the passive mechanical response

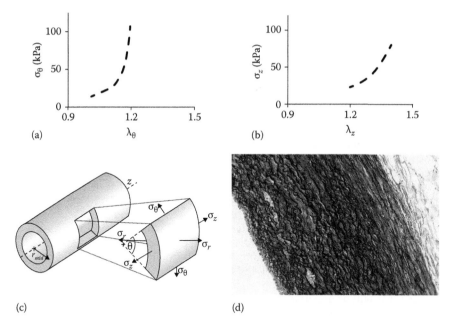

Figure 2.2 (a) Circumferential and (b) axial stress versus stretch of a porcine left anterior coronary artery. (c) The cylindrical coordinate system direction of stresses are shown in cartoon form. (d) Histological cross section of that artery using a combination Verhoeff–Masson's stain. (a and b: From Prim, D.A. et al., *J. Mech. Behav. Biomed. Mater.*, 54, 93, 2016.)

during the loading (pressurization or axial extension) of a left anterior descending coronary artery taken from a healthy adult pig (Figure 2.2). Defining stress σ and stretch λ in a convenient cylindrical coordinate system, we have for the circumferential direction θ the following two equations:

$$\sigma_\theta = \frac{Pr_i}{h}, \quad \lambda_\theta = \frac{r_{mid}}{R_{mid}}, \tag{2.1}$$

where
   P is the transmural pressure
   h is the deformed wall thickness
   r_o and r_i are inner and outer wall radii, respectively, in the deformed configuration
   $r_{mid} = (r_i + r_o)/2$ is the deformed midwall radius
   $R_{mid} = (R_i + R_o)/2$ is the reference mid-wall radius

The leftmost equation can be found by performing a force balance on a longitudinal section of the blood vessel. It represents the average transmural circumferential stress across the wall. Similarly in the axial z direction we have

$$\sigma_z = \frac{F}{\pi\left(r_o^2 - r_i^2\right)}, \quad \lambda_z = \frac{\ell}{L}, \tag{2.2}$$

where
   F is the axial force
   ℓ and L are the axial lengths of the vessel in the deformed and reference configurations, respectively

The stresses described by Equations 2.1 and 2.2 are defined in the current configuration and are called Cauchy stresses. Other stresses, and a general approach to the definition of stress, will be provided later in the chapter. The resultant data are shown in Figure 2.2 for the loading curves following a series of five preconditioning cycles. Observation of Figure 2.2a and b clearly indicates the nonlinear relationship between stress and stretch, while anisotropy is indicated by the differences between the stress–stretch relationships in these two directions (Figure 2.2c). Heterogeneity, on the other hand, was not measured explicitly in the mentioned study. However, a visualization of material heterogeneity is obvious from a stained cross section of the wall where the constituents of the tunica media (middle portion) differ from those of the tunica adventitia (right side - outer portion) (Figure 2.2d). In support of the concept of blood vessel mechanical heterogeneity, others have tested arterial layers separately for their mechanical properties and have observed marked mechanical differences (Holzapfel et al., 2000; von Maltzahn et al., 1981).

Clearly, soft biological materials have considerably more complex mechanical properties than those of traditional engineering materials such as metals and ceramics. Yet an assumption often invoked in error is that of linearity. Inspection of Figures 2.1 and 2.2 reveals that if a linear behavior is ascribed to a nonlinear material (e.g., a single tangential or Young's modulus) a large error will be generated for any point away from the tangent location. Research in the area of soft tissue mechanics over the last 50 years has provided the groundwork, with sufficient detail, to incorporate the earlier mentioned characteristics into the requisite mathematical models. Still, as is common with many engineering problems, a set of reasonable assumptions are used to simplify the problem.

One such assumption is that of incompressibility. Since most soft biological tissues in the human body consist largely of water (bulk modulus $\sim 2.15 \times 10^9\,Pa$), pressure that a living tissue could be exposed to would result in negligible volume changes when water is bound within the structure. Unfortunately, many tissues under consideration are porous in nature and readily exchange water, nutrients, and other fluids with the surrounding environment. In reconstituted collagen hydrogels, for example, the mechanical behavior is greatly dependent on the porosity of the hydrogel and alignment of the collagen fibers. For many soft tissues that have undergone preconditioning (e.g., the blood vessel shown in Figure 2.2), the incompressible assumption is a helpful one for determining three-dimensional strains. Still, careful biomechanical experimentation and the appropriate theoretical framework can be used to invoke or relax the assumption of incompressibility as needed.

The mechanical behavior of soft tissues is endowed by the cellular interactions with fibrous protein networks and an amorphous gel-like ground substance consisting of water, glycoproteins, glycosaminoglycans, and proteoglycans. Glycosaminoglycans such as hyaluronan, chondroitin sulfate, dermatan sulfate, and heparin sulfate, to name a few, are hydrophilic, unbranched polysaccharides. Mechanically, they serve as lubricants and shock absorbers in tissues while occupying a large volume per mass. Proteoglycans such as decorin, versican, syndecan, and aggrecan, to name a few, are glycosaminoglycans covalently attached to a protein core. Glycosaminoglycans and proteoglycans, through their affinity for water, contribute to a high resistance to compressing loading and add to the viscous behavior of soft tissues. Fibronectin and laminin are glycoproteins of the ECM that play a role in cell–cell and cell–substrate adhesion and have specific binding sites for other ECM components. Cells connect to ligands of the ECM through adhesion complexes that include transmembrane receptor proteins such as integrins (Griendling et al., 2011). Integrins link the intracellular stress fibers to the ECM enabling the transduction of extracellular signals into the cell. The acute mechanical influence of integrins and their ligands is most obvious in activated muscle cells that can shift the stress–strain curve.

In terms of fibrous protein networks, collagen and elastin are the main contributors to the mechanics of the ECM. Collagen is the most abundant protein in mammals—making up 25%–30% of all proteins in the body. It has a number of different isoforms, including the fibrillar and nonfibrillar types, of which collagen Type I makes up 90% of the body's collagen. As a fibrous protein, collagen contributes considerably to the degree of observed anisotropy. This behavior is demonstrated by a tendency for fibers to align in the direction of loading (e.g., predominant axial orientation of collagen fibers in tendons). Collagen is tensional stiff and stretch by only 10% once uncoiled. Elastic fibers, on the other hand, are responsible for large deformation and recoil and consist of microfibrillar (fibrillins and fibullins) and amorphous (elastin) components. Due to their desmosine and isodesmosine cross-links, elastin is biochemically stable and can stretch up to 150% of the original length (Ross and Pawlina, 2016). Whereas elastin tends to dominate the mechanical behavior at low stretches, collagen is the main contributor to nonlinear mechanics at higher stretches (Figure 2.1a).

To introduce and elucidate the relationships that define the nonlinear mechanics of soft biological materials in a concise manner, the equations shown throughout the current chapter are defined in the Cartesian coordinate system and in three-dimensional space wherever possible (base vectors $\mathbf{e}_x$, $\mathbf{e}_y$, and $\mathbf{e}_z$). For many soft biological materials (e.g., membranes, rectangular sections of cardiac tissue), this coordinate system is convenient; yet for tube-shaped tissues (e.g., blood vessels, lymphatics, airways), a set of cylindrical coordinates are preferred (see also Figure 2.2). Therefore, a vascular mechanics example is later presented using cylindrical coordinates.

2.2 DEFINITIONS OF STRESS

The proportional relationship between the force generated by the extension of an elastic spring can be described by Hooke's law. Force measurements alone, however, are not sufficient to analyze the mechanical behavior of materials. Instead, the concept of stress is used to describe how oriented forces act on oriented areas. In the simple case of uniaxial extension, where the applied force and the area over which it is applied are perpendicular, the average uniaxial normal stress is simply

$$\text{Stress} = \frac{\text{Force}}{\text{Area}}. \tag{2.3}$$

This formula results in units of Newtons divided by square meters (Pascals) in the international system. As most materials are stretched, intuitively the cross-sectional area will change. When the area used to calculate a stress refers to the current deformed cross-sectional area, it is called the Cauchy or True stress (Equations 2.1 and 2.2). In fact, there are a number of different types of stress that produce different results. Each type of stress has utility in certain situations of mechanical analysis. For example, in one-dimensional engineering, stress is convenient for materials undergoing small deformations where the cross-sectional area is easily measured a priori but not during loading. This approach has limited utility in soft tissue mechanics however and should be used sparingly. Regardless, to enable an accurate interpretation of reported results in the literature, it is essential to identify the form of stress (Table 2.1) used in any mechanical description or analysis (Ogden, 1997).

A value of stress is not unique, but rather depends on the orientation of applied force and the orientation of the area. Here we define a traction vector **t** that relates the differential force $\Delta\mathbf{f}$ to the oriented differential area Δa of the deformed configuration in three dimensions. Taken as the $\lim(\Delta a \to 0)$, we get Cauchy's postulate, which states

$$\mathbf{t}^{(\mathbf{n})} = \frac{d\mathbf{f}}{da}, \qquad (2.4)$$

where
  f is the force vector
  a is a scalar quantity for the deformed cross-sectional area
  $\mathbf{t}^{(\mathbf{n})}$ is a vector defined by the outward unit normal vector $\mathbf{n} = n_x\mathbf{e}_x + n_y\mathbf{e}_y + n_z\mathbf{e}_z$ acting on that surface

In the simplest case, when the outward unit normals are in the direction of the base vector (as shown in Figure 2.3), we get three vectors defined by the components:

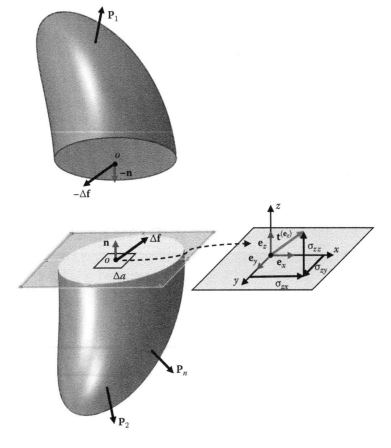

Figure 2.3   (Left) Applied arbitrary loads $\mathbf{P}_1$ to $\mathbf{P}_n$ deform a body. An infinitesimal resultant force vector $\Delta\mathbf{f}$ is defined at a point o on a fictitious plane through at body acting over a differential area Δa with unit normal vector **n**. (Right) A traction vector **t** relates the differential force to the differential area.

$$\mathbf{t}^{(\mathbf{e}_x)} = t_x^{(\mathbf{e}_x)}\mathbf{e}_x + t_y^{(\mathbf{e}_x)}\mathbf{e}_y + t_z^{(\mathbf{e}_x)}\mathbf{e}_z$$

$$\mathbf{t}^{(\mathbf{e}_y)} = t_x^{(\mathbf{e}_y)}\mathbf{e}_x + t_y^{(\mathbf{e}_y)}\mathbf{e}_y + t_z^{(\mathbf{e}_y)}\mathbf{e}_z \tag{2.5}$$

$$\mathbf{t}^{(\mathbf{e}_z)} = t_x^{(\mathbf{e}_z)}\mathbf{e}_x + t_y^{(\mathbf{e}_z)}\mathbf{e}_y + t_z^{(\mathbf{e}_z)}\mathbf{e}_z.$$

Cauchy's stress theorem then states that there exists a second-order tensor field defined in the deformed configuration associated with a plane with normal vector **n** so that

$$\mathbf{t}^{(\mathbf{n})} = \boldsymbol{\sigma} \cdot \mathbf{n}. \tag{2.6}$$

The lower-case notation is used here, and predominantly throughout the chapter, to indicate a quantity defined in the deformed configuration while capital letters are reserved for quantities defined in the reference configuration. Moreover, vector and tensoral quantities are shown in boldface. The nine components of the Cauchy stress tensor that completely define the state of stress at a point are thus

$$\boldsymbol{\sigma} = \begin{bmatrix} \sigma_{xx} & \sigma_{xy} & \sigma_{xz} \\ \sigma_{yx} & \sigma_{yy} & \sigma_{yz} \\ \sigma_{zx} & \sigma_{zy} & \sigma_{zz} \end{bmatrix}, \tag{2.7}$$

where the subscripts indicate the directions of stress associated with the oriented area (first subscript) and the direction of the applied force (second subscript). Stresses with repeating subscripts are normal stresses and those with nonrepeating subscripts are shear stresses. Furthermore, a material where the traction vector is zero across one of the surfaces is considered to be under plane stress. In those cases, the stress analysis is simplified since the stress tensor (but not the strain tensor) has a dimension of 2 instead of 3. Thin planar structures such as flat plates, or membranes with two dimensions much greater than the third, that are loaded in the directions parallel to the plate can be modeled using plane stress. Gently curved tissues such as thin-walled cylinders can sometimes also be modeled under the assumption that the radial stresses are negligible, thereby reducing the stress tensor down to only two dimensions.

It is often experimentally more convenient to measure the cross-sectional area A of a material prior to testing. By defining the traction vector **T** in the undeformed configuration with unit normal vector **N**, also in the undeformed configuration, we have a similar formulation that describes another measure of stress

$$\mathbf{T}^{(\mathbf{N})} = \frac{d\mathbf{f}}{dA} = \mathbf{P} \cdot \mathbf{N}, \tag{2.8}$$

where **P** is the first Piola–Kirchhoff stress, nominal stress, or multidimensional engineering stress. When deformations are small we can see that $a \approx A$, and these two stresses (Equations 2.6 and 2.8) are indistinguishable. Small deformations, however, are not typically of interest for soft biological materials and the differences between these two types of stresses can become substantial for materials of interest in this chapter. We will see in Section 2.5, however, that these two definitions of stress are interrelated through the deformation gradient **F**.

At every point in a stressed body, principal planes exist where there are no shear stresses, that is, the traction vector is purely normal and **t** is aligned parallel to **n**. As a result, the stress tensor has only diagonal components called

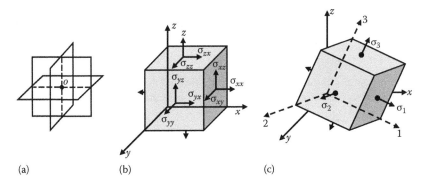

(a) (b) (c)

Figure 2.4 (a) Illustration of a material point o with six faces of stress in the Cartesian coordinate system. (b) The stresses acting on these six faces are better visualized as a cube where Cauchy stress is denoted by $\sigma_{(face, direction)}$. (c) The cube can be rotated in such a way so that shear forces are zero and only normal stresses exist. Such a rotation results in principal stresses $\sigma_1, \sigma_2, \sigma_3$.

principal stresses (Figure 2.4). Since the coordinate directions (x, y, z) no longer have meaning after the coordinate transformation to these principal directions, only a single subscript is used to distinguish the principal stress values from normal stress components and these are considered principal normal stresses. Careful design of many soft tissue experiments can be made to use a coordinate system with axes oriented in the principal directions thereby simplifying calculations. To find the principal stresses, μ is introduced as a constant of proportionality in (2.6) so that

$$(\sigma - \mu\mathbf{I}) \cdot \mathbf{n} = 0, \tag{2.9}$$

where $\mathbf{I}$ is the identity matrix. In component form it looks like

$$\begin{bmatrix} \sigma_{xx} - \mu & \sigma_{xy} & \sigma_{xz} \\ \sigma_{yx} & \sigma_{yy} - \mu & \sigma_{yz} \\ \sigma_{zx} & \sigma_{zy} & \sigma_{zz} - \mu \end{bmatrix} \begin{bmatrix} n_x \\ n_y \\ n_z \end{bmatrix} = \begin{bmatrix} 0 \\ 0 \\ 0 \end{bmatrix} \tag{2.10}$$

and only has a nontrivial solution if

$$|\sigma - \mu\mathbf{I}| = 0. \tag{2.11}$$

The three principal stresses σ_1, σ_2, and σ_3 are thus found by solving the characteristic equation (2.11) for the roots μ. It follows that $\sigma_1 \geq \sigma_2 \geq \sigma_3$ and the principal stresses are assembled into matrix form so that

$$\sigma = \begin{bmatrix} \sigma_1 & 0 & 0 \\ 0 & \sigma_2 & 0 \\ 0 & 0 & \sigma_3 \end{bmatrix}. \tag{2.12}$$

The resulting values are real numbers due to the symmetry of the stress tensor but do not depend on the coordinate system in which the components were initially given. When the value of a tensor is independent of the coordinate system, it is considered coordinate invariant. Expanding the determinant of (2.11), we can use the form

$$-\mu^3 + I_\sigma\mu^2 - II_\sigma\mu + III_\sigma = 0. \tag{2.13}$$

where the following scalar expressions I_σ, II_σ, and III_σ are defined

$$I_\sigma = tr(\sigma) = \sigma_{xx} + \sigma_{yy} + \sigma_{zz}$$

$$II_\sigma = \frac{1}{2}\left[tr(\sigma)^2 - tr(\sigma^2)\right]$$

$$= \sigma_{xx}\sigma_{yy} + \sigma_{yy}\sigma_{zz} + \sigma_{xx}\sigma_{zz} - \sigma_{xy}^2 - \sigma_{yz}^2 - \sigma_{zx}^2 \tag{2.14}$$

$$III_\sigma = det(\sigma)$$

$$= \sigma_{xx}\sigma_{yy}\sigma_{zz} + 2\sigma_{xy}\sigma_{yz}\sigma_{zx} - \sigma_{xy}^2\sigma_{zz} - \sigma_{yz}^2\sigma_{xx} - \sigma_{zx}^2\sigma_{yy}.$$

where $tr(\sigma)$ is the trace of σ and $det(\sigma)$ is the determinant of σ.

In the above equations, I_σ, II_σ, and III_σ are the first, second, and third principal invariants of stress. Although principal stresses are invariant, they should not be confused with the scalar quantities of the principal invariants (2.14). When the principal invariants are known, the principal stresses can be easily calculated from (2.13) and the three principal directions can be found from **n** in (2.9) or (2.10). The principal values and invariants of a tensor are simply properties of all tensors and we will see that these properties will also be used to describe the strain later in the chapter.

2.3 DEFINITIONS OF STRAIN

Deformations of a body are best described using the concept of strain, which relate two or more physical states of the same biological material after loading (Humphrey, 2002; Malvern, 1969). Using the Lagrangian description, we consider a time-invariant particle $P(X, Y)$ on material body in a reference (undeformed) configuration and that same particle in a deformed configuration $p(x, y, z)$. The position of that particle on the material body in the deformed configuration is indicated by the position vector $\mathbf{x} = \mathbf{x}(\mathbf{X})$ while the position of that same particle in the reference configuration using the same origin is indicated by a vector $\mathbf{X}$. The displacement vector of that particle is given by

$$\mathbf{u}(\mathbf{X}) = \mathbf{x}(\mathbf{X}) - \mathbf{X}, \tag{2.15}$$

and has components

$$\mathbf{u} = \begin{bmatrix} u_x \\ u_y \\ u_z \end{bmatrix} = \begin{bmatrix} x - X \\ y - Y \\ z - Z \end{bmatrix}. \tag{2.16}$$

The values of the displacement vector described by (2.15) and (2.16) can vary from point to point. Using the del operator ∇ to take the partial derivative of the displacement vector with respect to each material coordinate gives the displacement gradient tensor

$$\nabla \mathbf{u} = \begin{bmatrix} \dfrac{\partial u_x}{\partial X} & \dfrac{\partial u_x}{\partial Y} & \dfrac{\partial u_x}{\partial Z} \\[2ex] \dfrac{\partial u_y}{\partial X} & \dfrac{\partial u_y}{\partial Y} & \dfrac{\partial u_y}{\partial Z} \\[2ex] \dfrac{\partial u_z}{\partial X} & \dfrac{\partial u_z}{\partial Y} & \dfrac{\partial u_z}{\partial Z} \end{bmatrix}. \tag{2.17}$$

To understand how a body is stretched and rotated as it moves from the reference to deformed configuration, we further define an infinitesimal material

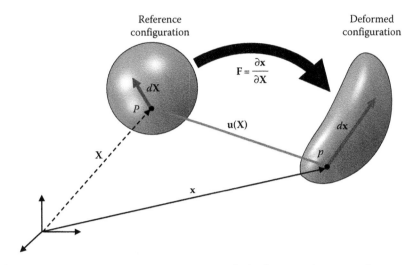

Figure 2.5 Diagrammatic representation of a body in a reference and deformed configurations. A displacement vector **u** is used to indicate the difference between the deformed **x** and reference position vectors **X** that map a particle P to p. An infinitesimal material vector in the reference configuration maps $d\mathbf{X}$ to $d\mathbf{x}$ through the deformation gradient **F**.

vector in the reference configuration as $d\mathbf{X}$ and after that material vector has been stretched and rotated as $d\mathbf{x}$ (Figure 2.5). The derivative of $d\mathbf{x}$ with respect to each component of $d\mathbf{X}$ gives an important quantity, the deformation gradient,

$$\mathbf{F} = \frac{\partial \mathbf{x}}{\partial \mathbf{X}} = \frac{\partial(x, y, z)}{\partial(X, Y, Z)} = \begin{bmatrix} \dfrac{\partial x}{\partial X} & \dfrac{\partial x}{\partial Y} & \dfrac{\partial x}{\partial Z} \\ \dfrac{\partial y}{\partial X} & \dfrac{\partial y}{\partial Y} & \dfrac{\partial y}{\partial Z} \\ \dfrac{\partial z}{\partial X} & \dfrac{\partial z}{\partial Y} & \dfrac{\partial z}{\partial Z} \end{bmatrix}. \tag{2.18}$$

F provides both deformation and rigid body motion of the material by relating the reference to the deformed configurations. The choice of a capital letter here is irrelevant since it spans both reference and deformed configurations, but is not to be confused with the lower-case notation for force. With (2.15) and (2.18) we have

$$\mathbf{F} = \frac{\partial}{\partial \mathbf{X}}(\mathbf{X} + \mathbf{u}) = \mathbf{I} + \frac{\partial \mathbf{u}}{\partial \mathbf{X}} = \mathbf{I} + \nabla \mathbf{u}. \tag{2.19}$$

Therefore, (2.18) can also be written in component form as

$$\mathbf{F} = \begin{bmatrix} 1 + \dfrac{\partial u_x}{\partial X} & \dfrac{\partial u_x}{\partial Y} & \dfrac{\partial u_x}{\partial Z} \\ \dfrac{\partial u_y}{\partial X} & 1 + \dfrac{\partial u_y}{\partial Y} & \dfrac{\partial u_y}{\partial Z} \\ \dfrac{\partial u_z}{\partial X} & \dfrac{\partial u_z}{\partial Y} & 1 + \dfrac{\partial u_z}{\partial Z} \end{bmatrix}. \tag{2.20}$$

Since stresses are related to deformations and not rigid body motion, we now introduce a symmetric tensor

$$\mathbf{C} = \mathbf{F}^T \cdot \mathbf{F}, \tag{2.21}$$

with the superscript T indicating the transpose of that tensor. This tensor, $\mathbf{C}$, is called the Right Cauchy–Green tensor and is defined in the reference coordinate system. Similarly, a Left Cauchy–Green tensor $\mathbf{B} = \mathbf{F} \cdot \mathbf{F}^T$ is defined in the deformed configuration and has utility in Cartesian coordinates. To allow strain to equal zero in the absence of deformation, we further define a Green or Lagrangian strain tensor as

$$\mathbf{E} = \frac{1}{2}(\mathbf{C} - \mathbf{I}). \tag{2.22}$$

From (2.22) the Green strain tensor is a 3×3 tensor with six independent components. Those components are

$$E_{XX} = \frac{\partial u_X}{\partial X} + \frac{1}{2}\left[\left(\frac{\partial u_X}{\partial X}\right)^2 + \left(\frac{\partial u_Y}{\partial X}\right)^2 + \left(\frac{\partial u_Z}{\partial X}\right)^2\right]$$

$$E_{YY} = \frac{\partial u_Y}{\partial Y} + \frac{1}{2}\left[\left(\frac{\partial u_X}{\partial Y}\right)^2 + \left(\frac{\partial u_Y}{\partial Y}\right)^2 + \left(\frac{\partial u_Z}{\partial Y}\right)^2\right]$$

$$E_{ZZ} = \frac{\partial u_Z}{\partial Z} + \frac{1}{2}\left[\left(\frac{\partial u_X}{\partial Z}\right)^2 + \left(\frac{\partial u_Y}{\partial Z}\right)^2 + \left(\frac{\partial u_Z}{\partial Z}\right)^2\right]$$

$$E_{XY} = E_{YX} = \frac{1}{2}\left(\frac{\partial u_X}{\partial Y} + \frac{\partial u_Y}{\partial X} + \frac{\partial u_X}{\partial X}\frac{\partial u_X}{\partial Y} + \frac{\partial u_Y}{\partial X}\frac{\partial u_Y}{\partial Y} + \frac{\partial u_Z}{\partial X}\frac{\partial u_Z}{\partial Y}\right)$$

$$E_{YZ} = E_{ZY} = \frac{1}{2}\left(\frac{\partial u_Y}{\partial Z} + \frac{\partial u_Z}{\partial Y} + \frac{\partial u_X}{\partial Y}\frac{\partial u_X}{\partial Z} + \frac{\partial u_Y}{\partial Y}\frac{\partial u_Y}{\partial Z} + \frac{\partial u_Z}{\partial Y}\frac{\partial u_Z}{\partial Z}\right)$$

$$E_{ZX} = E_{XZ} = \frac{1}{2}\left(\frac{\partial u_Z}{\partial X} + \frac{\partial u_X}{\partial Z} + \frac{\partial u_X}{\partial Z}\frac{\partial u_X}{\partial X} + \frac{\partial u_Y}{\partial Z}\frac{\partial u_Y}{\partial X} + \frac{\partial u_Z}{\partial Z}\frac{\partial u_Z}{\partial X}\right), \tag{2.23}$$

and are complete representations of strain. The nonlinear terms add complexity to analytical solutions so a simplification is introduced. For example, when deformations are small and the higher-order terms of (2.23) approach zero. The result is the linearized Green strain ε. Given in Cartesian coordinates, the components of ε are

$$\varepsilon_{xx} = \frac{\partial u_x}{\partial x}, \quad \varepsilon_{xy} = \varepsilon_{yx} = \frac{1}{2}\left(\frac{\partial u_x}{\partial y} + \frac{\partial u_y}{\partial x}\right)$$

$$\varepsilon_{yy} = \frac{\partial u_y}{\partial y}, \quad \varepsilon_{yz} = \varepsilon_{zy} = \frac{1}{2}\left(\frac{\partial u_y}{\partial z} + \frac{\partial u_z}{\partial y}\right) \tag{2.24}$$

$$\varepsilon_{zz} = \frac{\partial u_z}{\partial z}, \quad \varepsilon_{zx} = \varepsilon_{xz} = \frac{1}{2}\left(\frac{\partial u_z}{\partial x} + \frac{\partial u_x}{\partial z}\right),$$

Again, for infinitesimal elongations, Equations (2.23) and (2.24) are equivalent, but for finite deformations they are not. Since the current chapter is focused on the behavior of soft biological materials that are characterized by finite deformations, Equation 2.24 has limited utility. Likewise, the deformation gradient is related to the linearized Green strain by

$$\varepsilon = \frac{1}{2}\left(\mathbf{F} + \mathbf{F}^T - 2\mathbf{I}\right).$$
(2.25)

Like stresses, strain invariants that are independent of the coordinate system can also be found. The invariant equations are the same for all measures of strain (e.g., linearized and nonlinearized Green strain) but are shown here for the left Cauchy–Green strain tensor $\mathbf{C}$, which will be used in the examples of later sections. The three invariants of $\mathbf{C}$ are thus

$$
\begin{aligned}
I_C &= tr(\mathbf{C}) \\
II_C &= \frac{1}{2}\left[tr(\mathbf{C})^2 - tr(\mathbf{C}^2)\right], \\
III_C &= \det(\mathbf{C}),
\end{aligned}
$$
(2.26)

where I_C, II_C, and III_C are the first, second, and third invariants of the right Cauchy–Green strain tensor. The principal strains can be calculated in a similar manner to the principal stresses and are not included here for brevity.

2.4 MECHANICAL EQUILIBRIUM

Newton's second law states that the sum of the forces acting on a particle is equal to the mass times the acceleration of that particle. In continuum mechanics, that concept is extended and generalized by the momentum balance equations so that the forces acting on the surface of a body or on the body itself must be balanced. In tensor notation, this is written as (Humphrey, 2002)

$$\nabla \cdot \boldsymbol{\sigma} + \rho \mathbf{b} = \rho \mathbf{a},$$
(2.27)

where
ρ is the density
$\mathbf{b}$ and $\mathbf{a}$ are the body force and acceleration vectors, respectively

The components in the Cartesian coordinate system for a full 3-D state of stress are

$$
\begin{aligned}
\frac{\partial \sigma_{xx}}{\partial x} + \frac{\partial \sigma_{yx}}{\partial y} + \frac{\partial \sigma_{zx}}{\partial z} + \rho g_x &= \rho a_x \\
\frac{\partial \sigma_{xy}}{\partial x} + \frac{\partial \sigma_{yy}}{\partial y} + \frac{\partial \sigma_{zy}}{\partial z} + \rho g_y &= \rho a_y \\
\frac{\partial \sigma_{xz}}{\partial x} + \frac{\partial \sigma_{yz}}{\partial y} + \frac{\partial \sigma_{zz}}{\partial z} + \rho g_z &= \rho a_z,
\end{aligned}
$$
(2.28)

with the only body forces present due to gravity g. Often we neglect gravity (i.e., $g_x = g_y = g_z = 0$) and consider the quasi-static problem (i.e., $a_x = a_y = a_z = 0$) to reduce the number of terms in (2.28). For reference, equilibrium equations in cylindrical coordinates will be given for an example problem later in this chapter. For certain cases in mechanics, the equilibrium equations can be solved without introducing constitutive relations and therefore are independent of the material (cf., (2.1) through (2.3)). Such solutions are considered universal (Humphrey, 2002).

2.5 CONSTITUTIVE RELATIONS

In mechanics, constitutive relations are the basis for identifying material mechanical properties and insomuch describe the relationships between stresses and strains (Fung, 1993; Humphrey, 2002). These properties depend on the constitution of the material and the conditions of which the material is tested (e.g., temperature, loading rate). A historically popular constitutive relation is for an elastic solid that follows Hooke's law where stress is linearly proportional to strain. This law further assumes that the material is homogeneous and isotropic. As a result, these materials can be completely described by only a handful of constants (e.g., Young's modulus and Poisson's ratio). From the restrictions described earlier in the chapter, we can see that this behavior cannot be ascribed to soft biological materials that generally display nonlinearity, elasticity, anisotropy, and heterogeneity. Careful experimentation and adherence to principles of continuum mechanics can be used, however, to simplify the constitutive relations and find adequate descriptors of the nonlinear mechanics of soft biological materials.

From the second law of thermodynamics for an isothermal problem, the general constitutive relations are based on the derivative of a scalar-valued strain energy function W so that

$$\mathbf{S} = \frac{\partial W}{\partial \mathbf{E}} = 2\frac{\partial W}{\partial \mathbf{C}}, \tag{2.29}$$

where $\mathbf{S}$ is the second Piola–Kirchhoff stress. This particular description of stress has a convoluted physical meaning. Instead, the utility of $\mathbf{S}$ lies in the mathematical convenience of Equation 2.29. Regardless, through the deformation gradient, and its determinant, all three measures of stress (Table 2.2) (i.e., Cauchy and the first and second Piola–Kirchhoff stresses) can be calculated from each other via

$$\sigma = \frac{1}{\det \mathbf{F}}\mathbf{F} \cdot \mathbf{S} \cdot \mathbf{F}^T = \frac{1}{\det \mathbf{F}}\mathbf{F} \cdot \mathbf{P}. \tag{2.30}$$

Therefore, it is also true that the first and second Piola–Kirchoff stresses are related through $\mathbf{P} = \mathbf{S} \cdot \mathbf{F}^T$. The general form for the Cauchy stress tensor can then be defined as

$$\sigma = \frac{1}{\det \mathbf{F}}2\mathbf{F}\frac{\partial W}{\partial \mathbf{C}}\mathbf{F}^T = \frac{1}{\det \mathbf{F}}\mathbf{F}\frac{\partial W}{\partial \mathbf{E}}\mathbf{F}^T. \tag{2.31}$$

Equation 2.31 is shown here with respect to the right Cauchy–Green ($\mathbf{C}$) or Green ($\mathbf{E}$) strain tensors, respectively. For an incompressible material modeled under finite elasticity, Equation 2.31 is further modified so that

$$\sigma = -p\mathbf{I} + 2\mathbf{F}\frac{\partial W}{\partial \mathbf{C}} \cdot \mathbf{F}^T = -p\mathbf{I} + \mathbf{F}\frac{\partial W}{\partial \mathbf{E}} \cdot \mathbf{F}^T, \tag{2.32}$$

where p is a scalar quantity called the Lagrange multiplier and is used to enforce incompressibility. Equilibrium and boundary conditions are used to find the value of the Lagrange multiplier as demonstrated in the representative example at the end of this chapter.

2.5.1 Strain Energy Functions

As discussed earlier, finite elasticity is an idealized mechanical behavior of a material in which strain magnitudes are large, the loading and unloading paths are either coincident or assumed to be so, and stress–strain states are independent of time (i.e., are instantly realized upon loading/unloading). The development of continuum-based constitutive models in the framework of finite elasticity can be achieved via identification of a strain energy function (SEF) W. If this methodology is successfully applied, meaning an analytical form for a SEF is proposed and parameters are identified, the modeled behavior is termed hyperelastic. The identification of an SEF for constitutive modeling of soft materials is of indisputable utility and a cornerstone of biological materials engineering.

The SEF is a scalar function that in the most general sense returns the strain energy stored per unit volume of a material based on the deformation gradient, that is, $W = W(\mathbf{F})$. In accordance with the theory of finite elasticity, the strain energy stored within the material is fully recoverable and the deformation process is reversible. If the deformation is isothermal, the stress associated with a specific deformed state can be calculated based on the SEF, and thus stress–strain relations can be derived. The development of constitutive stress–strain relations enables accurate prediction of biological material behavior only under conditions of interest sufficiently close to those in which they were initially defined. Clearly, an extreme change in environment, for example, a temperature difference or an alteration of the material hydration state, may significantly impact mechanical behavior and necessitate a distinct constitutive formulation. For this reason, it is important that experimental testing for the purpose of SEF identification adequately reflects the conditions in which a given material will be deployed.

Assuming the applicability of an identified SEF for a given biological material and environmental conditions, subsequent predictions of the mechanical response and quantifications of stress and strain provide critical information to evaluate performance criteria and understand the behavior of resident mechanosensitive cells. In regard to the latter, mechanosensitive cells incorporated within a material will sense and respond to their immediate mechanical environment, which is defined by local stresses and strains that are calculable as opposed to measurable quantities. Therefore, the identification of an SEF is an essential step toward understanding both biomechanical and mechanobiological phenomena of soft biological materials.

To enable the calculation of stresses, the effects of rigid body motion that are nominally contained in $\mathbf{F}$ must be eliminated. To this end, W is typically expressed as a function of $\mathbf{C}$ or $\mathbf{E}$, that is, $W = W(\mathbf{C})$ or $W = W(\mathbf{E})$. In some cases, these SEFs can be further simplified via expression as a function of the strain tensor invariants as opposed to the generalized strain tensors (cf., Equation 2.26). For example, if a material is mechanically isotropic, then W can be expressed as $W = W(I_C, II_C, III_C)$. If in addition to exhibiting isotropy, the material is incompressible, then by definition $III_C = 1$ and $W = W(I_C, II_C)$. These particular simplifications are useful in the development of mechanical models for a wide range of soft materials that have no preferred orientation of load-bearing constituents and contain a high volume fraction of water. Examples include the disorganized collagen matrices present in scar tissue and the materials categorized as amorphous hydrogels.

2.5.2 Classification of Strain Energy Functions

Many constitutive models that quantify the nonlinear finite elastic behavior of soft biological materials assume both mechanical incompressibility and anisotropy. As mentioned earlier, incompressibility is usually a valid assumption due to the high water content of these materials, and is thereby a consequence of both bulk chemical properties and the fact that biological materials are typically immersed within well-hydrated environments, that is, in vivo. The preponderance of mechanically anisotropic models, particularly for natural materials, can be interpreted as evidence of soft tissue adaptation to a specific loading modality, and therefore constitutes a mechanical example of form following function. In many soft tissues, such as the arterial wall, relatively stiff fibers within an ECM are aligned in such a manner to adequately support the applied loads. Such microstructural alignment underlies anisotropic material properties when viewed on the subcontinuum scale. If alternatively these tissues were to possess a microstructure with no preferred fiber orientation, they might still effectively bear loads albeit in a less efficient manner. Clearly, these characteristics (incompressibility and anisotropy) are not required in the development of constitutive models of nonlinear finite elastic behavior, but are emphasized here due to their particular relevance to soft biological materials.

Given a set of descriptive characteristics of material behavior, constitutive models are proposed and identified based on the judicious selection of an SEF with a specific functional form. One could reasonably divide constitutive models proposed for soft biological materials into three main groups. *Phenomenological models* employ an SEF in terms of a single analytical function of the components of the strain tensor, from which all stress–strain constitutive relations follow. The functional form and material parameters of the SEF are identified based on a best fit between experimentally recorded data and corresponding theoretical predictions. In some important historical cases, an educated guess based on mathematical convenience and/or desired mechanical response characteristics has been successfully used to propose a functional form even prior to data acquisition. For example, the well-known Mooney–Rivlin model developed in the 1940s for rubberlike elasticity was proposed based on the desire to yield nonlinear stress–strain behavior under tensile loads and linear relationships in shear (Mooney, 1940). The resultant framework is remarkably successful in modeling the behavior of elastomers over large strain ranges, and is still used today. While phenomenological models can accurately match experimental data, they cannot offer insight into the structural and compositional features of a biological material that give rise to observed mechanical behavior. Conversely, *structure-motivated models* account for major structural load-bearing constituents of biological materials, for example, elastin and collagen, by using functional forms that attempt to delineate their individual contributions. In these models, the identified values of material-specific parameters can provide insight into the relative importance of key structural components, but only to the extent in which their effects are captured by the proposed analytical form. Finally, *structural-based models coupled with the theory of constrained mixtures* utilize SEFs that account for the individual mechanical properties and the amount of key structural load-bearing constituents. Here, the incorporation of compositional information into the SEF allows for further decoupling of key component roles in governing the overall mechanical behavior.

The selection of a particular class of SEF, and moreover a specific analytical form, should be guided by the envisioned applications of the resultant constitutive framework. For illustrative purposes, select examples of constitutive models proposed for both compressible and incompressible isotropic elastic materials are tabulated here (Table 2.3). For these and any proposed SEFs, particular attention should be given to the bounding values of material parameters and the initial assumptions about operative material behavior.

Anisotropic forms of SEFs are also commonly used in soft tissue biomechanics. For example, the fully orthotropic version of the Fung exponential shown in Table 2.3 is given by

$$W = c\left(e^{Q}-1\right),$$
$$Q = c_1 E_{11}^2 + c_2 E_{22}^2 + c_3 E_{33}^2 + 2c_4 E_{11}E_{22} + 2c_5 E_{22}E_{33} + 2c_6 E_{33}E_{11} \qquad (2.33)$$
$$+ 2c_7\left(E_{12}^2 + E_{21}^2\right) + 2c_8\left(E_{23}^2 + E_{32}^2\right) + 2c_9\left(E_{31}^2 + E_{13}^2\right),$$

Table 2.3: Examples of SEFs Proposed for Incompressible and Compressible Isotropic Elasticity

Name	Functional Form	Material Parameters	Operative Material Behavior
Neo-Hookean	$W(I_C) = c_1(I_C - 3)$	c_1	Incompressible isotropic elasticity
Mooney–Rivlin	$W(I_C, II_C) = c_1(I_C - 3) + c_2(II_C - 3)$	c_1, c_2	Incompressible isotropic elasticity
Rivlin–Sawyers	$W(I_C, II_C, III_C) = c_1(I_C - 3) + c_2(II_C - 3) + c_3(II_C - 3)^2$	c_1, c_2, c_3	Incompressible isotropic elasticity
Fong–Penn	$W(I_C, II_C, III_C) = c_1(I_C - 3) + c_2(II_C - 3)$ $+ f(I_C, II_C)(III_C - 1) + \dfrac{1}{8K}(III_C - 1)^2$	c_1, c_2, K	Compressible isotropic elasticity
Blatz Ko	$W(I_C, II_C, III_C) = \dfrac{\mu\alpha}{2}\left(I_C - 1 - \nu^{-1} + (1 - 2\nu)III_C^{-\nu/(1-2\nu)}\nu^{-1}\right)$ $+ \dfrac{\mu(1-\alpha)}{2}\left(\dfrac{II_C}{III_C} - 1 - \nu^{-1} + (1 - 2\nu)III_C^{\nu/(1-2\nu)}\nu^{-1}\right)$	μ, α, ν	Compressible isotropic elasticity
Fung exponential (isotropic form)	$W(I_E, II_E) = c(e^{Q_{iso}} - 1)$ $Q_{iso} = \alpha I_E^2 + \beta II_E$	c, μ, β	Incompressible isotropic elasticity

where c_{1-9} are material parameters. The Fung exponential is an excellent fit for most soft biological materials and can be reduced down to simpler forms depending on the level of isotropy and/or observed strain behavior (e.g., as shown in Table 2.3).

A structurally motivated strain energy function that incorporates both the isotropic and anisotropic fibrous contributions was first described by Holzapfel et al. (2000) and later refined by others to investigate multiple fiber families (Baek et al., 2007). This SEF is colloquially referred to as the HGO model and has proven to be an excellent structurally motivated SEF for vascular mechanics as it allows for fibers to be preferentially aligned. In this SEF, the isotropic contribution is given as a neo-Hookean material while the anisotropic part is represented by an exponential function

$$W_{iso} = b_0 \left(I_C - 3 \right),$$

$$W_{anis} = \sum_{k=1-4} \frac{b_{k1}}{2b_{k2}} \left\{ \exp\left[b_{k2} \left(\sqrt{\lambda_\theta^2 \sin^2 \alpha_k + \lambda_z^2 \cos^2 \alpha_k} - 1 \right)^2 \right] - 1 \right\}, \quad (2.34)$$

where k represents a family of fibers (in this case 4) that point in the direction of the angle α. b_{k1} and b_{k2} are material parameters for these fiber families while b_0 is the isotropic material parameter. A structurally based SEF can easily be formulated from this structurally motivated SEF using mass fractions that can be measured experimentally (e.g., through microscopy) of constituents that contribute to the isotropic or specific families of fiber orientations,

$$W = \phi_{iso} W_{iso} + \sum_{k=1-4} \phi_k W_{anis}. \quad (2.35)$$

with ϕ representing the mass fraction of the constituent that contributes to the isotropic or anisotropic response. Often, but not exclusively, the isotropic contribution is attributed to elastin while the anisotropic fiber families are associated with collagen or circumferentially acting smooth muscle cells (Ferruzzi et al., 2013).

2.6 MATERIAL PARAMETER IDENTIFICATION

Following the selection of a particular functional form for the SEF, the development of a constitutive formulation requires the identification of material-specific parameter values. In general, this is accomplished via nonlinear regression techniques that seek parameter values that provide the best fit between theoretically predicted and experimentally recorded mechanical quantities (e.g., pressure, force, stress). Numerous regression techniques have been successfully used for this purpose, and available software packages have significantly simplified what was once a mathematically challenging task. However, the accuracy of obtained mechanical data, which can be diminished by nonuniform sample geometry, measurement errors, and the extent to which the test setup reflects the intended loading scenario, will inherently impact the accuracy of identified material parameters. Unfortunately, experimental mechanical tests involving soft biological materials present significant challenges, including proper stabilization of samples within the testing device and account of the local and confounding effects at the sample-grip interface, and as a result, obtained data sets typically contain significant error. Obviously, repeat measurements, high data acquisition rates, and large sample numbers should be used to circumvent testing challenges and provide the most accurate data set for model development.

Therefore while the identification of material parameters based on the processing of experimental data is now a straightforward task, the resultant constitutive model will only be as accurate as the mechanical data used for its identification.

2.7 CONSTITUTIVE MODEL VALIDATION

Validation is a key aspect in the development of all physical models, including those that describe and predict material mechanical behavior. To extend beyond the descriptive value, a model must yield predictions about material behavior under conditions that differ from those used to identify parameters. In the most rigorous sense, a constitutive mechanical model is only applicable under the conditions used for its identification. Even this may not capture all of the intricacies of the mechanical behavior of that tissue. Therefore a conundrum exists, wherein the researcher must carefully select a test (and resultant data set) for validation in which the identified constitutive model will retain a predictive capacity and that model must be reasonably close to the expected behavior. Indeed, one objective in mechanical modeling is the specification of conditions under which the identified constitutive relations are valid, that is, are capable of yielding predictions that quantitatively agree with recordable mechanical behavior.

Constitutive model validation is not always straightforward for biological materials, as the identified parameters typically exhibit such large sample-to-sample variation that accurately predicting untested sample behavior is not always possible. Nevertheless, there are at least two basic approaches for model validation. For one, a given sample can be subjected to distinct mechanical tests (say uniaxial compression and uniaxial extension), with one data set (say tension) used for parameter identification and the other for model validation. Clearly, this is a viable approach only if the constitutive formulation is applicable to both types of deformation. Data obtained from multiaxial tests can be allocated in a similar manner, meaning that some are used for parameter identification and some reserved for validation. The second approach could be using the constitutive model to solve complex boundary value problems with geometries and boundary conditions that are markedly different than those used for parameter identification. While analytical solutions to geometrically complex problems are typically not available, commercial software that employs finite element methods have proven particularly useful for this purpose.

2.8 UTILITY OF CONSTITUTIVE FORMULATIONS

2.8.1 Quantification of Mechanical Properties

Identification of a constitutive formulation for a biological material, or any material, implies that mechanical properties have been quantified. Identified model parameters that are in fact intrinsic mechanical properties reflect the constitution of the material, albeit within the context of finite elasticity and the employed functional form. The specific class of constitutive model (phenomenological, structure motivated, or structure based) will dictate the interpretation of parameter meaning and determine the extent to which parameter values can be related to the compositional and/or structural features of the material. Clearly, with phenomenological models, structure–property relations are difficult to establish, whereas structure-motivated/based models may offer this insight. Irrespective of model class, the comparative values of identified parameters in different experimental groups can promote understanding of how certain variables impact material mechanical properties. For example, comparative studies could be designed to quantify how sample age, chemical/mechanical conditioning, exposure to bioactive compounds, cell/tissue source, or disease states affect the mechanical properties of a given type of biological material. As mentioned earlier, significant sample-to-sample variability in identified parameter values is expected for biological materials, typically necessitating well-controlled experiments and relatively large sample sizes to discern average mechanical properties of a specific material type.

2.8.2 Boundary Value Problems

An added advantage of developing constitutive models is the ability to solve boundary value problems of interest in biological materials engineering. Both direct and inverse boundary value problems can be formulated following the identification of a constitutive model. A *direct* boundary value problem is one in which in addition to the constitutive model, the geometry, loads, boundary conditions, evolution equations, and hypothesis about the nature of the material response are given. The solution to these problems includes a quantification of the material mechanical response and therefore forms the basis for stress and strain analyses. Direct boundary value problems have significant relevance to both biomechanics and mechanobiology, and are essential for quantifying the local mechanical environment of mechanosensitive cells within a biological material.

In contrast, a number of *inverse* boundary value problems can also be solved, following the identification of a constitutive model. For example, if a targeted mechanical response is known for given loads and boundary conditions, an inverse problem can be formulated in which the solution entails quantification of the requisite sample geometry. In these problems, the solution can provide insight into the optimal design of structure based on the targeted mechanical response. Similarly, if instead the geometry is known and the loads are unknown, then the obtained solution can establish the admissible loading conditions of the structure based on the targeted mechanical response.

Although conceptually straightforward, very few analytical solutions to boundary value problems in nonlinear mechanics exist. Instead, numerical solutions and powerful computational approaches including finite elements must be used to obtain solutions.

2.8.3 Computational Modeling

A major application of an identified constitutive model for biological materials is thus the development of computational simulations of mechanical behavior. Of most relevance are computational models based on finite element approaches. Although these models are a manifestation of boundary value problems, they are reiterated here for the purpose of further describing their particular utility in biological materials engineering.

When sample geometries and/or loading of materials undergoing nonlinear elastic deformations extend beyond a select few cases (i.e., uniaxial extension of a strip or inflation–extension of a tube), analytical solutions are unavailable. In these cases, computational models that employ finite element methods offer powerful tools for solving both direct and inverse boundary value problems of interest. Numerous commercially available finite element packages have streamlined the incorporation of computational simulation into engineering practice and have resulted in a marked expansion of these techniques across the research, development, and regulatory phases of biological materials. In these packages, geometries can be easily built using integrated computer-aided design (CAD) modules or imported/reconstructed from other sources. The spatial discretization of these geometries into area (2-D models) or volume (3-D models) elements is likewise facilitated by software packages, and requires specification of the desired degree of discretization (mesh density). Following assignment of material properties, applied loads, and boundary conditions, execution of an underlying mathematical scheme in the context of finite elements is a heavily automated process and may require little user knowledge of the employed method for obtaining a solution.

Clearly, the ease by which computational simulations can be performed with available technology has both advantages and disadvantages. On the one hand, the accessibility of this technique to a broad range of researchers has accelerated the development of numerous biological materials and increased the fundamental value of identified constitutive models. However, presented simulation results must be cautiously interpreted, as the degree of automation now characteristic in computational simulation inevitably results in some misapplication of this technique. Nevertheless, there is no question that the development of computational simulations based on constitutive models of nonlinear elastic behavior will continue to be a major avenue of fundamental research and product development with regard to soft biological materials.

2.9 REPRESENTATIVE EXAMPLE: INFLATION AND EXTENSION OF A LEFT ANTERIOR DESCENDING PORCINE CORONARY ARTERY

Here we present an example of a common experiment in vascular mechanics, the inflation and extension of a blood vessel. The results are from the set of sample data presented in Section 2.1 (Prim et al., 2016). Here the vessel is initially in an unloaded configuration and then inflated by pressure and axially extended. All data are collected following five preconditioning cycles. The geometry of this experiment is greatly simplified by using a cylindrical coordinate system and the motion at any point in this material is described by

$$r = r(R), \quad \theta = \chi\Theta, \quad z = \lambda Z, \tag{2.36}$$

where r, θ, z are the radii, circumference, and axial lengths, respectively, in the deformed configurations with reference dimensions described by R, Θ, Z. χ is related to an opening angle that helps account for residual stresses. For many materials, a stress can exist even in the absence of external loads. These stresses are called residual stresses and serve to homogenize the stress field in arteries. The deformation gradient for the motion described in (2.36) is

$$\mathbf{F} = \begin{bmatrix} \dfrac{\partial r}{\partial R} & \dfrac{1}{R}\dfrac{\partial r}{\partial \Theta} & \dfrac{\partial r}{\partial Z} \\ r\dfrac{\partial \theta}{\partial R} & \dfrac{r}{R}\dfrac{\partial \theta}{\partial \Theta} & r\dfrac{\partial \theta}{\partial Z} \\ \dfrac{\partial z}{\partial R} & \dfrac{1}{R}\dfrac{\partial z}{\partial \Theta} & \dfrac{\partial z}{\partial Z} \end{bmatrix} = \begin{bmatrix} \dfrac{\partial r}{\partial R} & 0 & 0 \\ 0 & \chi\dfrac{r}{R} & 0 \\ 0 & 0 & \lambda_z \end{bmatrix}, \quad \chi = \dfrac{\pi}{\pi - \Phi}. \tag{2.37}$$

Now assuming that the blood vessel is incompressible, we let $\det \mathbf{F} = 1$ so that

$$\frac{\partial r}{\partial R}\chi\frac{r}{R}\lambda_z = 1. \tag{2.38}$$

Integration from the inner wall yields an expression for the form of $r = r(R)$ in (2.36) so that

$$r = \sqrt{\frac{1}{\chi \lambda_z}\left(R^2 - R_i^2\right) + r_i^2}\,. \tag{2.39}$$

From (2.21) and (2.22) the right Cauchy–Green strain tensor is

$$\mathbf{C} = \begin{bmatrix} \left(\dfrac{\partial r}{\partial R}\right)^2 & 0 & 0 \\ 0 & \left(\chi \dfrac{r}{R}\right)^2 & 0 \\ 0 & 0 & \lambda_z^2 \end{bmatrix} = \begin{bmatrix} \left(\dfrac{1}{\chi \lambda_\theta \lambda_z}\right)^2 & 0 & 0 \\ 0 & \left(\chi \lambda_\theta\right)^2 & 0 \\ 0 & 0 & \lambda_z^2 \end{bmatrix}. \tag{2.40}$$

Therefore from (2.32) $\boldsymbol{\sigma} = -p\mathbf{I} + 2\mathbf{F} \cdot (\partial W / \partial \mathbf{C}) \cdot \mathbf{F}^T$ and using the chain rule we can get

$$2\frac{\partial W}{\partial \mathbf{C}}\mathbf{F}^T = \frac{\partial W}{\partial \mathbf{F}^T}\,.$$

The nonzero components of the Cauchy stress tensor are thus

$$\sigma_r = -p + \lambda_r \frac{\partial W}{\partial \lambda_r}, \quad \sigma_\theta = -p + \lambda_\theta \frac{\partial W}{\partial \lambda_\theta}, \quad \sigma_z = -p + \lambda_z \frac{\partial W}{\partial \lambda_z}. \tag{2.41}$$

Equilibrium in the absence of body forces and in cylindrical coordinates is represented by

$$\frac{\partial \sigma_{rr}}{\partial r} + \frac{1}{r}\frac{\partial \sigma_{\theta r}}{\partial \theta} + \frac{\partial \sigma_{zr}}{\partial z} + \frac{\sigma_{rr} - \sigma_{\theta\theta}}{r} + \rho g_r = 0$$

$$\frac{\partial \sigma_{r\theta}}{\partial r} + \frac{1}{r}\frac{\partial \sigma_{\theta\theta}}{\partial \theta} + \frac{\partial \sigma_{z\theta}}{\partial z} + 2\frac{\sigma_{r\theta}}{r} + \rho g_\theta = 0 \tag{2.42}$$

$$\frac{\partial \sigma_{rz}}{\partial r} + \frac{1}{r}\frac{\partial \sigma_{\theta z}}{\partial \theta} + \frac{\partial \sigma_{zz}}{\partial z} + 2\frac{\sigma_{rz}}{r} + \rho g_z = 0.$$

with the only nontrivial equations given as

$$\frac{d\sigma_{rr}}{dr} = \frac{\sigma_{\theta\theta} - \sigma_{rr}}{r} = 0, \quad -\frac{1}{r}\frac{\partial \sigma_{\theta\theta}}{\partial \theta} \rightarrow \frac{\partial p}{\partial \theta} = 0, \quad \frac{\partial \sigma_{zz}}{\partial z} \rightarrow \frac{\partial p}{\partial z} = 0. \tag{2.43}$$

From (2.43), we see that that the Lagrange multiplier $p = p(r)$ but $p \neq p(\theta, z)$. Integration of (2.43) with limits of r_i to r and substituting (2.41) gives an equation that is solved for the Lagrange multiplier

$$p(r) = P - \int_{r_i}^{r} \left(\lambda_\theta \frac{\partial W}{\partial \lambda_\theta} + \lambda_r \frac{\partial W}{\partial \lambda_r}\right)\frac{dr}{r} + \lambda_r \frac{\partial W}{\partial \lambda_r}\,. \tag{2.44}$$

Equation 2.44 allows for the calculation of the complete stress field when the kinematics are known and the functional form of W chosen. Additionally,

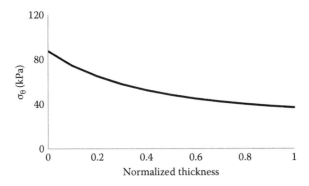

Figure 2.6 The transmural distribution of stresses in the Porcine Left Anterior Descending Artery at 100 mmHg. The mass fractions of the isotropic (attributed to elastin) and anisotropic (attributed to collagen) were found to be $\phi_{iso} = 0.14$ and $\phi_c = 0.34$ (see also Figure 2.2). The material parameters found through nonlinear regression to be $b_0 = 3.47$ kPa, $b_{11} = 18.7$ kPa, $b_{12} = 1.81$, $b_{21} = 1.96$ kPa, $b_{22} = 3.15$ kPa, $b_{31} = 7.51$ kPa, $b_{32} = 40.5°$ with opening angle $\Phi = 70.1°$ and stress-free thickness $H = 0.88$ mm. (Data from Prim, D.A. et al., *J. Mech. Behav. Biomed. Mater.*, 54, 93, 2016.)

the transmural pressure can be calculated by integration limits between r_i and r_o in (2.43) so that

$$P = \int_{r_i}^{r_o} \left(\lambda_\theta \frac{\partial W}{\partial \lambda_\theta} + \lambda_r \frac{\partial W}{\partial \lambda_r} \right) \frac{dr}{r}. \tag{2.45}$$

Moreover, applying a mean axial boundary condition, the force can be calculated from

$$F = \int_{r_i}^{r_o} \lambda_z \frac{\partial W}{\partial \lambda_z} r \, dr - \pi r_i^2 P. \tag{2.46}$$

Thus far, the approach has been independent of W, but did rely on the incompressibility assumption. Now we apply the anisotropic, structurally based HGO strain energy function represented by Equations 2.34 through 2.35. The material parameters associated with the constitutive model were determined via the nonlinear regression of equation of the error between the experimental equations (2.1) and (2.2) and modeling equations (2.45) and (2.46) using a least-squares technique. The lower and upper limits of the parameters were prescribed as b_0 and $b_{k1} \in [0, 105]$, $b_{k2} \in [0, 10]$, and $\alpha \in [0°, 90°]$, while the mass fractions of ϕ_{iso} and ϕ_{anis} were determined through light microscopy and a thresholding analysis (e.g., Figure 2.2d). The resulting stress distribution through the wall of the blood vessel is shown in Figure 2.6.

2.10 REMARKS

Endowed by a dynamic composite microstructure consisting of complicated interactions between cells, proteins, and water-dominated ground substances, the characteristic mechanical behavior common to most soft biological materials has been presented. Here the finite strain theory of continuum mechanics has been introduced for the analysis of such behavior. The theoretical framework, constitutive formulations, and strategies for parameter identification are also delineated while many key assumptions employed to generate tractable solutions to boundary value problems are acknowledged. A clinically relevant example, using data taken from a left anterior coronary artery in our lab, is shown to demonstrate the types of data and analysis common to soft tissue mechanics. Overall, the materials and strategies presented herein are not designed to be all encompassing, but instead to provide a jumping-off point for interested scientists, engineers, and clinicians to formulate more advanced studies in soft tissue mechanics. Concepts of the mechanical behavior of soft biological materials have been around for centuries but implementation of the engineering approach, technological advancements, and continual revision of existing achievements by creative researchers have enabled an improved understanding of the behavior of soft biological materials. As the biomedical field evolves further, mechanics will continue to play a pivotal role in disease progression and interventional strategies.

REFERENCES

Baek, S., Gleason, R.L., Rajagopal, K.R., Humphrey, J.D., 2007. Theory of small on large: Potential utility in computations of fluid–solid interactions in arteries. *Comput. Methods Appl. Mech. Eng.* 196, 3070–3078.

Ferruzzi, J., Bersi, M.R., Humphrey, J.D., 2013. Biomechanical phenotyping of central arteries in health and disease: Advantages of and methods for murine models. *Ann. Biomed. Eng.* 41, 1311–1330.

Fung, Y.C., 1993. *The Mechanical Properties of Living Tissues*, Springer-Verlag, New York.

Griendling, K.K., Harrison, D.G., Alexander, R.W., 2011. Chapter 8. Biology of the vessel wall. In: Fuster, V., Walsh, R.A., Harrington, R.A. (Eds.), *Hurst's the Heart, 13e*, The McGraw-Hill Companies, New York.

Holzapfel, G., Gasser, T., Ogden, R., 2000. A new constitutive framework for arterial wall mechanics and a comparative study of material models. *J. Elast. Phys. Sci. Solids* 61, 1–48.

Humphrey, J.D., 2002. *Cardiovascular Solid Mechanics: Cells, Tissues, and Organs*, Springer, New York.

Humphrey, J.D., Delange, S.L., 2004. *An Introduction to Biomechanics: Solids and Fluids, Analysis and Design*, Springer, New York.

Malvern, L., 1969. *Introduction to the Mechanics of a Continuous Medium SE*, Prentice-Hall Series in Engineering of the Physical Sciences, Prentice Hall, Englewood Cliffs, NJ.

Mooney, M., 1940. A theory of large elastic deformation. *J. Appl. Phys.* 11, 582–592.

Ogden, R.W., 1997. *Non-Linear Elastic Deformations*, Dover Civil and Mechanical Engineering, Dover Publications, Mineola, NY.

Park, J., Lakes, R.S., 2007. *Biomaterials, Biomaterials: An Introduction*, 3rd edn. Springer, New York.

Prim, D.A., Zhou, B., Hartstone-Rose, A., Uline, M.J., Shazly, T., Eberth, J.F., 2016. A mechanical argument for the differential performance of coronary artery grafts. *J. Mech. Behav. Biomed. Mater.* 54, 93–105.

Ross, M., Pawlina, W., 2016. *Histology: A Text and Atlas*, 7th edn., Wolters Kluwer Health, Philadelphia, PA, pp. 156–171.

Snodgrass, M.E., 2015. *World Clothing and Fashion: An Encyclopedia of History, Culture, and Social Influence*, Taylor & Francis, London.

von Maltzahn, W.-W., Besdo, D., Wiemer, W., 1981. Elastic properties of arteries: A nonlinear two-layer cylindrical model. *J. Biomech.* 14, 389–397.

PART II

BIOMATERIALS IN DEVICES AND MEDICINE

3 Biomaterials in Devices

Danieli C. Rodrigues, Izabelle M. Gindri, Sathyanarayanan Sridhar, Lucas Rodriguez, and Shant Aghyarian

CONTENTS

3.1 INTRODUCTION

The field of biomaterials has experienced tremendous growth since its introduction more than half a century ago as researchers progressed in understanding biocompatibility, the reaction of the body to foreign objects, and the mechanical environment of the body. Since then, there has been steady progress in the development of new biomaterials and implantable devices for a variety of applications and branches [1,2]. Today, there is a biomaterial to augment or replace each part of the body integrating different synthetic and/or biologically inspired materials. Traditionally, a biomaterial was defined as "a nonviable material used in a medical device intended to interact with biological systems" [1]. Other classical definitions followed including the requirement for inertness and structural stability. According to Williams, such ambiguous usage evolved over time by restricting the use of the word "biomaterials" to situations in which "a material exerts beneficial effects on human health by specific, direct, and intentional contact with tissues and where the nature of the interaction is important in determining the performance of a device" [3]. With an increasing degree of sophistication, biomaterials today not only have applications in device design, but they also incorporate biological active components and dynamic behavior [4], giving them crucial roles in diagnostics, biological screening, drug delivery, and tissue engineering applications [4].

In this chapter, a brief review of the use of biomaterials in devices is presented with emphasis on developmental aspects in the field, general requirements for biomaterials in implant design, classes of biomaterials, and selected implant applications (orthopedic and dental) with focus on mechanical requirements and performance.

3.1.1 Evolution in the Development of Biomaterials and Implantable Devices

Early biomaterials included gold for dental use, iron to rejoin fractured femurs, wood for limb prostheses, animal bone as needles, and glass to replace eyes [1,4,5]. Selection of materials was based on availability and functionality. The biggest challenge was finding materials or structures that could match the host tissues in terms of mechanical properties [6]. However, host response to these materials varied depending on the physician technique and the condition of host tissues. Such variability motivated the eventual use of synthetic polymers, metallic alloys, and ceramics in place of naturally derived materials due to their superior mechanical stability and consistency [4].

Developments in the biomaterials field over the last 65 years can be divided into three distinct generations [1]. The first generation of biomaterials (1950–1970) focused on bioinertness as the primary requirement in order to minimize reactions with the human body. The goal of the materials from the first generation was to achieve a combination of functionalities that could match the properties of host tissues; examples of materials used include stainless steel, titanium alloys, poly(methyl methacrylate) (PMMA), and polyethylene (PE). The second generation of biomaterials (1970–1990) focused on the synthesis of bioactive and resorbable materials, with rates of degradation that could be tailored to match specific applications. By achieving bonding with tissues or tuned degradation into soluble and nontoxic products, the biomaterial–implant interface could be eliminated, which in turn would prevent foreign body reactions with host tissues. Examples of biomaterials developed in the second generation included bioceramics (hydroxyapatite, tricalcium phosphate), degradable polymers (poly(lactic-co-glycolic acid), polylactic acid [PLA], polycaprolactone [PCL]), and natural polymers (collagen). The third generation of biomaterials (1990–current) aims to achieve the regeneration of functional tissues (regenerative medicine and tissue engineering) by true replacement with living cells. Biomaterials from this generation can be biointeractive, integrative, resorbable, and can stimulate specific cell responses. Examples of biomaterials from the third generation include composites, degradable polymers for scaffolds and drug delivery, hydrogels, biological molecules, collagen, etc.

Advances through these three generations resulted in important improvements in the properties and biological interactions of biomaterials, which in turn increased the life span and performance of devices such as orthopedic implants, vascular stents, contact lenses, dental materials, and neural devices [4]. The devices available today, with a variety of material combinations and geometries, are fine-tuned to better match patient anatomy and clinical condition.

3.1.2 General Requirements and Attributes for Biomaterials in Implant Design

Requirements for implantable biomaterials vary with each application due to the various mechanical, chemical, and biological environments found in the human body. The first requirement for the selection of a biomaterial is its compatibility within the biological system [3,7]. The introduction of biomaterials into a living tissue is known to trigger a series of host responses leading to wound healing. The host response consists of different phases such as tissue damage, homeostasis (blood clotting), inflammatory phase, change in ionic concentration, neutrophil infiltration, phagocytic effect of monocytes and macrophages, angiogenesis, proliferative phase, repair and remodeling, formation of granular fibroblast/connective tissue, and finally scar formation [1].

Besides biocompatibility, the success of a biomaterial is also dependent on its properties [1,3,6]. Some of the most important material properties evaluated when assessing minimum performance requirements for biomaterials and implants are yield strength, modulus of elasticity or stiffness, hardness, fatigue strength, wear resistance, and corrosion. These parameters can provide information on the mechanical response of materials under different loading scenarios, endurance limits, and disposition to particle generation due to mechanical (wear) or electrochemical processes (corrosion). In addition to material properties, the health status of a patient and the surgical technique used to implant the material can also affect implant performance.

In summary, a biomaterial or implantable device is expected to (1) perform the desired function for the intended duration, (2) cause no harm to the patient, and (3) be delivered in an efficient and safe manner [3]. Table 3.1 summarizes the general attributes in the selection of biomaterials for implant design. Functionality is dependent on the application. For example, a metal that exhibits a combination of high strength and high stiffness while being used in the design of articulation

Table 3.1: General Attributes in the Selection of Biomaterials and Implant Design

	General Features	Surface Characteristics, Geometry
Functionality	Mechanical	Mechanically strong to withstand mechanical environment. Combination of appropriate ductility, yield strength, modulus of elasticity, hardness, fracture resistance, toughness, resilience, creep, fatigue strength. Resistant to wear, friction, fretting
	Chemical	Nontoxic and noncarcinogenic. Resistant to corrosion and degradation. Appropriate surface chemistry
	Physical	Appropriate electrical, thermal, optical, and magnetic properties. Specific permeability and lubricity
	Biological	Controlled biodegradation. Control of molecular targeting and cell phenotype. Does not trigger specific biological responses
Safety	Intrinsic biocompatibility	Appropriate local host response, absence of remote or systemic adverse effects
	Toxicology	Noncarcinogenic, nontoxic to host tissues (unless intentionally tuned to specific targets such as cancers and bacterial biofilms), nonallergenic, blood compatible, noninflammatory
Development processes	Manufacturing	Machinable, moldable, extrudable
	Sterilization	Withstand conditions of commercial sterilization techniques
	Regulatory	Compliance with federal and/or international regulations for design, development, and commercialization. Regulatory regimes vary by country

Sources: Williams, D., *Essential Biomaterials Science*, 1st edn., Cambridge University Press, Cambridge, U.K., 2014; Patel, N.R. and Gohil, P.P., *Int. J. Emerg. Technol. Adv. Eng.*, 2, 91, 2012.

components (e.g., cobalt–chromium [CoCr] alloys) may not be appropriate for bone-contacting interfaces. This is because of elastic modulus or modulus mismatch between the bone and CoCr interface that can lead to stress shielding. The biomaterial must also satisfy its design functions while in service, for example, an orthopedic implant transmits loads and distributes stresses to the surrounding tissues, implanted lenses focus on and transmit light, and a pacemaker provides electrical stimuli [7]. Safety is a mandatory attribute to ensure appropriate local and systemic host response and control of cytotoxicity. Finally, development processes are dependent on the type of biomaterial. That is, manufacturing, sterilization, and packaging procedures and the design features of the implant in consideration will dictate, in part, the quality and regulatory routes.

3.1.3 Classes of Implantable Materials Used in Devices

According to Williams, medical devices encompass a range of structural implants, drug and gene delivery systems, tissue engineering, organ printing, cell patterning, microelectronic devices, imaging, and diagnostic products [8]. These devices are typically composed of metals, synthetic or natural polymers, ceramics, composites, nanoparticles, and quantum dots.

The kind of biomaterial to be selected for an implantable medical device is dependent on the required function for each application. Biocompatibility, porosity, degradation profile, and mechanical properties are some of the parameters that can drive the selection of a specific material. Table 3.2 summarizes the general characteristics of the most common classes of materials used today in implantable devices. Metals, ceramics, and polymers are used individually or in combination in a range of biomedical applications.

In terms of structural applications, the mechanical characteristics of the tissue to be replaced need to be taken into consideration in the selection of the

Table 3.2: General Characteristics of Main Classes of Biomaterials Used in Implant Design

Material	General Characteristics	Examples of Use
Metals	Excellent electrical and thermal conductivity, appropriate mechanical properties (strength, stiffness, ductility, and toughness), biocompatibility and corrosion resistance, reasonable cost, amenable to surface treatments	Joint replacements, fracture fixation components, oral and maxillofacial reconstructions, cardiovascular interventions, sutures, screws, surgical instruments
Polymers	Resilient, easy to manufacture through solution casting, melt molding or machining techniques. Polymers can be made inert, biodegradable, or reactive	Sutures, adhesives, soft tissues, orthopedic implants, dental materials, tissue scaffolds, drug delivery, wound dressings, blood vessels, vascular grafts, breast implants, contact lenses, intraocular lenses, coatings, bone cements, catheters, kidney, bladder
Ceramics	Bioinert, resorbable, bioactive, high compressive strength, good tribological properties (bioinert type)	Dental implants, dental crowns, dental bridges, femoral heads, bone screws and plates, fillers, porous coatings

Sources: Ratner, B.D. et al., *Biomaterials Science—An Introduction to Materials in Medicine*, 3rd edn., Elsevier, Waltham, MA, 2013; Williams, D., *Essential Biomaterials Science*, 1st edn., Cambridge University Press, Cambridge, U.K., 2014; Agrawal, C.M. et al., *Introduction to Biomaterials: Basic Theory with Engineering Applications*, 1st edn., Cambridge University Press, Cambridge, U.K., 2014.

Table 3.3: Bulk Mechanical Properties of Common Biological Materials

Tissue	Young's Modulus E (GPa)	Tensile Strength σ_{UTS} (MPa)
Hard tissue		
Bone		
Cortical bone, longitudinal	11–21	60–150
Cortical bone, transverse	5–13	50
Cancellous bone	0.05–5	10–20
Teeth		
Tooth dentin	11	40
Tooth enamel	83	10
Soft tissue		
Cartilage, articulation	11	28
Ligament	300	30
Tendon	402	47
Intraocular lenses	6	2
Skin	0.1–0.2	8
Arterial wall	0.001	0.50–1.72

Sources: Ratner, B.D. et al., *Biomaterials Science—An Introduction to Materials in Medicine*, 3rd edn., Elsevier, Waltham, MA, 2013; Williams, D., *Essential Biomaterials Science*, 1st edn., Cambridge University Press, Cambridge, U.K., 2014; Patel, N.R. and Gohil, P.P., *Int. J. Emerg. Technol. Adv. Eng.*, 2, 91, 2012; Black, J. and Hastings, G., *Handbook of Biomaterials Properties*, 1st edn., Springer, Berlin, Germany, 1998.

appropriate class of biomaterials. This will ensure better distribution of loads at the tissue–material interface, minimizing stress shielding effects. Stress shielding is characterized by reduction in bone density (osteopenia) as a result of removal of normal stress from the bone by an implant [9]. For example, metals and bioinert ceramics are typically selected for hard tissue replacement while polymers are more suitable for soft tissue applications. Table 3.3 illustrates the modulus of elasticity and tensile strength of examples of soft and hard tissues commonly treated with the implantable materials described in Table 3.2.

3.1.3.1 Metals

Metals constitute a large portion of the medical device industry. Typically, most implants or surgical instrumentation will have one or more metallic components [1,3,6]. Some examples of key applications of metallic alloys are illustrated in Table 3.4. The most common metals used in the design of these devices are titanium and its alloys, stainless steels, and CoCr alloys. Other metals such as tantalum (i.e., Trabecular Metal™) and magnesium alloys (degradable metal) are undergoing further developments due to their particular properties and potential to serve as alternative metals.

The way a metal is processed has a direct impact on its microstructure and purity, which in turn affects the material properties and performance. Although a metal in its pure state will not typically possess properties and activity desired, the material can be further processed via alloying, heat treatments, and surface modification to achieve improved properties [11]. Metals have a combination of strength, ductility, and toughness that make them the preferred choice in the design of load-bearing implants and internal fixation devices. Another

Table 3.4: General Applications and Properties of Common Metallic Biomaterials

Metal	Applications	Young's Modulus E (GPa)	Tensile Strength σ_{UTS} (MPa)
Pure titanium (cpTi)	Dental implants, plates for maxillofacial fixation	103	240–550
Ti6Al4V	Joint replacement, bone fixation devices, dental implants, nails and plates for orthopedic augmentation, heart valves	110	860
Nitinol (Ni–Ti)	Guide wires, stents, orthodontic arch wires, endodontic reamers and files	70–110 (austenitic) 21–69 (martensitic)	100–800 (austenitic) 50–300 (martensitic)
CoCr alloys	Bone plates for fracture fixation, screws, dental implants, stents, articulation components in hip and knee prostheses, surgical tools	210–253	655–1896
Stainless steels	Surgical instruments, spinal instruments, joint replacement, bone plate and screws for fracture fixation, heart valve	190	586–1351

Sources: Ratner, B.D. et al., *Biomaterials Science—An Introduction to Materials in Medicine*, 3rd edn., Elsevier, Waltham, MA, 2013; Agrawal, C.M. et al., *Introduction to Biomaterials: Basic Theory with Engineering Applications*, 1st edn., Cambridge University Press, Cambridge, U.K., 2014; Patel, N.R. and Gohil, P.P., *Int. J. Emerg. Technol. Adv. Eng.*, 2, 91, 2012.

advantage is that the most commonly employed metallic biomaterials (titanium and its alloys, CoCr alloys, and stainless steels) have a passive surface oxide layer that enhances their corrosion resistance in the biological environment. This passive oxide layer naturally forms on the surface of these metals in contact with oxygen from biological fluids or air. Oxide layers create a barrier to corrosion by protecting the metal surface against accelerated metal ion dissolution. This oxide layer also gives the surface of the metal roughness, which will encourage bone growth and osseointegration [12]. Thus, the presence of the oxide layer is key to the biocompatibility of metals.

3.1.3.1.1 Titanium and Titanium Alloys

Titanium is known for its light weight (density of 4.5 g/cm³), excellent corrosion resistance, and enhanced biocompatibility. Titanium and its alloys provide excellent biological, chemical, and mechanical properties compared to 316 stainless steel and CoCr alloy. This material has been vastly employed in the design of bone-contacting devices with a moderate modulus of elasticity of approximately 100–110 GPa [1,7], which closely resembles the elastic modulus of bone (1–10 GPa) [9,13]. The American Society for Testing and Materials (ASTM) International has classified commercially pure titanium (unalloyed form, cpTi) into four different grades (Grade 1–Grade 4). These classifications are based on the amount of impurities present (in particular oxygen, nitrogen, and iron), which have a marked effect on the material's properties. The alloy titanium–6 aluminum–4 vanadium (Ti6Al4V) is broadly employed in the medical device industry. Besides, the attributes already discussed, the Ti6Al4V alloy exhibits good workability, weldability, and heat treatment stability. Investigations of the long-term performance of Ti6Al4V implants have raised concerns with regard to potential adverse reactions caused by the release of vanadium [11,14] *in vivo*. Because titanium is relatively soft, mechanical (wear, fretting), chemical (corrosion), and synergistic degradation processes (tribocorrosion, fretting-crevice corrosion) can lead to metal ion and metallic particles' generation *in vivo*, which can cause adverse biological reactions [15,16]. In an attempt to solve issues associated with vanadium, other alloys have been developed such as the titanium–6 aluminum–7 niobium (Ti6Al7Nb) alloy, which has been employed in the fabrication of femoral hip stems and fracture fixation devices [9]. Nitinol, which belongs to the class of shape memory alloys, has a near-equiatomic composition of titanium and nickel (49%–51% Ni–Ti). This alloy is broadly employed in the design of self-expanding vascular stents, which can be plastically deformed at low temperature and return to their original pre-deformed shape when exposed to elevated temperatures. This property is a result of the martensitic–austenitic transformation this material undergoes.

3.1.3.1.2 Cobalt–Chromium Alloys

The alloy CoCr is the hardest, strongest, most fatigue-resistant of the alloys used in biomedical applications. CoCr alloys exhibit excellent wear properties and corrosion resistance. Due to these properties, this material is ideal for the design of components that undergo articulation. The four types of CoCr alloys commonly used in biomedical applications are ASTM F75 (Co–28Cr–6Mo casting alloy), ASTM F799 (Co–28Cr–6Mo thermodynamically processed alloy), ASTM F90 (Co–20Cr–15W–10Ni wrought alloy), and ASTM F562 (Co–35Ni–20Cr–10Mo wrought alloy) [6]. Although these alloys are similar in composition, the different processing methods used to make them can result in unique properties. Corrosion resistance is a result of high bulk chromium content and chromium oxide (Cr_2O_3). Cobalt-based alloys are highly resistant to corrosion even in severely acidic environments. Thermal treatments used for microstructural modification and surface treatments can significantly affect the electrochemical and mechanical properties of this material. Furthermore, CoCr alloys are extremely difficult to machine into the complex geometries of some devices by conventional machining methods. Although this alloy has excellent tribological properties, when used in the design of metal-on-metal articulation interfaces, it is known to cause severe wear problems with the release of Cr and Ni ions, which have been associated with the development of metallosis and pseudotumor formation. Metal-on-metal implants (CoCrMo acetabular cups and CoCrMo femoral heads) have been recalled from the U.S. market due to a number of patients presented with significant problems, which leads to revision surgeries and several lawsuits against orthopedic device companies [17,18].

3.1.3.1.3 Stainless Steel

Stainless steel alloys have been used in a wide range of applications due to easy availability, lower cost, good corrosion resistance, excellent fabrication properties, acceptable biocompatibility, and strength. Stainless steel alloys are iron-based alloys that contain at least 10.5% of chromium. The corrosion resistance of stainless steel is attributed to the formation of chromium oxides (Cr_2O_3) and can be improved by increasing the Cr content. Other alloying elements such as molybdenum and nickel can also be used to improve the material's mechanical and other physical properties. Based on microstructure, stainless steel alloys are classified as martensitic, ferritic, austenitic, and duplex alloys. Each of these materials has characteristic properties and varying chromium content (10.5%–30%) [6]. Austenitic stainless steel is extensively used for the design of medical implants and devices, in particular the 316 L stainless steel with low carbon content (<0.03%).

3.1.3.1.4 Recent Innovations

Tantalum (Trabecular Metal™) with a porosity of 80% has properties and characteristics that resemble trabecular bone and can be used in bone-contacting interfaces of orthopedic and dental implants. Magnesium is another material that has been currently explored due to its biodegradability and potential use as a temporary device. Using biodegradable materials in applications that need to provide structural support can be challenging, as their degradation rates must be controlled and slow enough that fracture stabilization is not endangered prior to healing. In order to control degradation, magnesium has been employed as alloys, magnesium–calcium (Mg–Ca) being common [19]. Other uses of biodegradable magnesium include scaffold linings of implants to enhance bone interaction with the surface, hence improving osseointegration [19].

3.1.3.2 Polymers

Polymers represent the largest class of materials used in medicine. The vast majority of which are carbon based [1]. This class of material possesses an array of tunable properties that make it applicable in the design of medical devices, drug delivery systems, and tissue engineering. A variety of natural and synthetic polymers have been used as implantable biomaterials. These can be made inert, biodegradable, or reactive. Table 3.5 provides examples of polymers used as implantable materials. Some of the unique properties of polymers include flexibility, good biocompatibility, a wide variety of compositions available

Table 3.5: General Applications and Properties of Examples of Polymeric Biomaterials

Polymer	Applications	Young's Modulus E (GPa)	Tensile Strength σ_{UTS} (MPa)
UHMWPE	Articulation components in orthopedic, spinal, and extremities implants	1–2	>30
PMMA	Bone cements, glass lenses, intraocular lenses, dentures, cosmetic surgery	1.8–3.1	45–75
Polypropylene	Sutures, repair of abdominal wall defects, films	1–1.6	28–36
PLA	Sutures and suture reinforcements, membranes for dentistry, scaffolds for tissue reconstruction, orthopedic fixation devices, skin replacement materials	1.2–3	28–50
PCL	Long-term drug delivery, degradable staple, sutures, implantable contraceptive device, stents	1.2	10–16

Sources: Ratner, B.D. et al., *Biomaterials Science—An Introduction to Materials in Medicine*, 3rd edn., Elsevier, Waltham, MA, 2013; Agrawal, C.M. et al., *Introduction to Biomaterials: Basic Theory with Engineering Applications*, 1st edn., Cambridge University Press, Cambridge, U.K., 2014; Patel, N.R. and Gohil, P.P., *Int. J. Emerg. Technol. Adv. Eng.*, 2, 91, 2012.

Note: Properties can vary with polymer molecular weight, manufacturing technique, and composition.

with well-characterized physical and mechanical properties, and ease of manufacturing into products with the desired shape and characteristics. Polymers can be easily shaped using solution casting, melt molding, and machining. They can also be functionalized with specific chemical groups to be more compatible with the application of interest. For example, polymers can be functionalized with antimicrobial compounds for increased resistance to biofilm formation.

Although advantageous, polymers are not as strong or stiff as metals or ceramics and therefore may not be the correct choice when an implant or biomaterial is required as load-bearing systems in function. The mechanical properties of polymers depend on their composition, structure of molecular chains, and molecular weight.

3.1.3.2.1 Inert Polymers

Inert polymers are used in load-bearing and non-load-bearing applications as permanent replacement. Examples of major inert polymers used as biomaterials are summarized in Table 3.5. Poly(methyl methacrylate), PMMA, has long clinical history in orthopedics, ophthalmology, and dentistry [20]. Combining its inertness with good mechanical properties, PMMA is an advantageous material because it can be prepared and manipulated in the clinical setting being delivered while in a viscous state (e.g., orthopedic bone cements), which undergoes polymerization on site, or as acrylic materials (e.g. dentures). In total joint replacement (TJR), bone cements provide additional augmentation in patients with low bone quality, thus the success of these procedures can depend on the performance of the cement used. Another important polymer system in biomaterials is polyethylene, PE, in particular ultrahigh-molecular-weight polyethylene (UHMWPE). This material has been vastly employed in articulation systems of total joints, spinal, and extremities' devices due to its good tribological properties. UHMWPE has over 40 years of successful clinical history in hip and knee implants [21,22]. PE has good creep resistance and high yield, which minimizes the potential for plastic deformation. The predominant problem presented in metal–polymer systems is the production of wear particles. UHMWPE can produce submicron and nano-sized wear particles in large quantities, which can trigger osteolysis and consequent implant loss [1,6]. Oxidation of UHMWPE is another aggravating problem and can occur if the material is sterilized in air with gamma radiation. Cross-linking with gamma radiation improves the wear resistance of the material but can negatively impact some of the mechanical properties of the material affecting its shelf-life due to oxidation. In recent years, cross-linking followed by post-radiation processing and antioxidant (vitamin E) incorporation is typically conducted to circumvent oxidation in UHMWPE [23].

3.1.3.2.2 Degradable Polymers

Degradable polymers are used as temporary load-bearing/non-load-bearing materials in drug delivery, temporary support devices, multifunctional implants, and tissue engineering scaffolds [1,3,6,24]. Degradable polymers are made up of monomers connected through functional groups with unstable links in the backbone, which are biologically degraded or eroded by enzymes or nonenzymatic processes. They can be of natural (collagen, albumin, gelatin, cellulose, casein, starch) or synthetic origin (aliphatic polyesters, polyphosphoesters, polyamino acids, etc.). Degradable polymers are advantageous because they circumvent the need for surgical removal and problems related to long-term stability, foreign body response, inflammation, and rejection [3,6]. However, these polymers can degrade into toxic products, undergo premature failure, and offer limited mechanical integrity, restricting their application in load-bearing systems. Examples of commonly employed degradable polymers and their properties are summarized in Table 3.5. Among the natural and synthetic systems approved for use *in vivo* PLA, poly(glycolic acid), and their copolymers are the most investigated and employed, being the system of choice in a number of medical applications.

3.1.3.2.3 Hydrogels

Hydrogels have gained attention in the last few years due to their high water absorbency and potential for applications in the medical field, particularly in tissue engineering [1,3,6]. They are composed by hydrophilic water-insoluble natural or synthetic polymers, which are cross-linked forming 3D networks. Some of their applications in the biomedical field include contact lenses (e.g., poly(2-hydroxyethyl methacrylate)), drug delivery systems (e.g., poly(N-2-hydroxypropyl); poly(2-hydroxyethyl methacrylate-co-methyl acrylate)), tissue engineering scaffolds (several combinations), wound healing, artificial tendon, articular cartilage, bioadhesives, and artificial skin.

3.1.3.3 Ceramics

Ceramics are refractory polycrystalline materials employed in many applications in the biomedical field due to a combination of attractive mechanical properties and biocompatibility [25]. Ceramics are inorganic, nonmetallic materials composed of two or more metallic and nonmetallic elements bond by ionic and covalent bonds. The structural arrangement of ceramics results in inertness, hardness and brittleness, high compressive strength, high corrosion resistance, excellent tribological properties (low wear and low coefficient of friction), and good thermal properties [1,6,25]. In the medical industry, ceramics have been used for a relatively long time in applications such as eye glasses, diagnostic instruments, thermometers, dentistry materials (porcelain crowns, glass-filled cements, and dentures), dental restoration materials, and medical sensors.

Ceramics are prepared by the action of heat and subsequent cooling, which have a marked effect on the material's properties. As a result, ceramics can have crystalline, partially crystalline, or semicrystalline, and amorphous structures. In terms of mechanical performance, ceramics typically fail with little plastic deformation and are very sensitive to the presence of flaws in their structure. Ceramics withstand high compressive loads prior to fracture; however, they perform poorly in tension because cracks propagate quite quickly leading to breakage.

Another important characteristic of ceramics is that they are made up of elements that are naturally present in tissues such as bone, which is composed of calcium phosphate (CaP); therefore, it is an interesting scaffold material [26]. Ceramics are classified as bioinert, bioactive, and resorbable. Examples of bioceramics used in the medical field include zirconia (ZrO_2), alumina (Al_2O_3), pyrolytic carbon, bioglass, glass ceramics hydroxyapatite, and tricalcium phosphate. Table 3.6 provides applications for these materials along with information on their mechanical properties.

3.1.3.3.1 Bioinert Ceramics

Bioinert ceramics are highly dense and smooth exhibiting a combination of great wear and corrosion resistance with high strength. Such properties qualify these materials for load-bearing applications in orthopedics and dentistry.

Table 3.6: General Applications and Properties of a Few Examples of Ceramic Biomaterials

Ceramic	Applications	Young's Modulus E (GPa)	Tensile Strength σ_{UTS} (MPa)
Alumina	Femoral heads and acetabular cups in total hip implants	380	350
Zirconia	Femoral heads in total hip implants, dental implants	150–200	200–500
Bioglass	Bioglass 45S5 (bone regeneration), implants in non-load-bearing areas, Bioglass 8625 (encapsulation of implanted devices)	20–35	56–83
Calcium phosphates	Powders, fillers, cements, coatings	40–117	69–193

Sources: Ratner, B.D. et al., *Biomaterials Science—An Introduction to Materials in Medicine*, 3rd edn., Elsevier, Waltham, MA, 2013; Williams, D., *Essential Biomaterials Science*, 1st edn., Cambridge University Press, Cambridge, U.K., 2014; Patel, N.R. and Gohil, P.P., *Int. J. Emerg. Technol. Adv. Eng.*, 2, 91, 2012.

Alumina and ZrO_2 have found applications in the biomedical field (Table 3.6). Alumina was in fact the first ceramic used clinically [7]. The first generation of ceramics in load-bearing applications showed problems associated with osteolysis *in vivo*, loosening of the acetabular component in hip implants, and possibility of ceramic fracture. These problems were related to the large grain-sized material that resulted from long sintering times [27,28]. The second generation of ceramics brought reduction in grain size and in the incidence of ceramic fractures through sintering treatments [28]. Unlike the first and second generations, the third generation of ceramics has improved mechanical strength due to processes such as isostatic pressing, laser etching, and proof testing [27,28]. Pyrolytic carbon is another interesting implantable material due to its good compatibility with bone, although it has limited load-bearing applications because of its intrinsic brittleness. Because bioinert ceramics only interface with host tissues with minimal bone ingrowth, movement at the biomaterial–tissue interface can occur.

3.1.3.3.2 Bioactive Ceramics

Bioactive ceramics are materials that promote biological fixation by forming chemical bonds with bone and soft tissues. Examples of materials in this group include bioglass and glass ceramics (e.g., $Na_2OCaOP_2O_3$-SiO), and some phases of calcium phosphate (e.g., hydroxyapatite sintered a high temperature). These materials undergo slow rate of degradation, if any present, and induce bone formation, which can promote extended implant life [29].

3.1.3.3.3 Resorbable Ceramics

Resorbable ceramics are materials that degrade over time to be replaced by natural tissue. Different phases of CaP ceramics such as hydroxyapatite ($Ca_{10}(PO_4)_6(OH)_2$) and tricalcium phosphate ($Ca_3(PO_4)_2$) are used as biomaterials with varying degradation rates [20]. CaP is naturally occurring in the body, for example, hydroxyapatite being the mineral component of bone. Although CaP has bone-like porosity and characteristics, its poor mechanical behavior can restrict its use as load-bearing material. CaP ceramics are used as porous coatings, dental fillers, cements, and reinforcing materials [26].

3.2 APPLICATIONS OF BIOMATERIALS IN ORTHOPEDICS

Biomedical devices used in the orthopedic field are mostly employed in joint replacement or fracture management procedures [9,13]. Joint replacement materials are used in hip, knee, shoulder, wrist, elbow, finger, and ankle arthroplasty procedures (Figure 3.1) while fracture management devices include plates, screws, pins, and bone cements. Although there is a wide range of materials that are used to design these devices, a particular set of metals, polymers, composites, and ceramics is normally used as discussed earlier. Understanding the mechanical properties of each material, as well as the interaction of these materials in each system and with living systems, is essential to comprehend gaps that still affect implant performance. For that, there is a series of experiments, which include standard tests regulated by the ASTM and the International Organization for Standardization (ISO), that need to be performed to characterize materials and devices in order to ensure safety, efficacy, and performance.

This section will focus on the biomaterials that are used to design orthopedic devices as well as the main required tests that need to be performed prior to commercialization. TJR systems, which include designs for total hip and knee replacement and devices employed in fracture management, will be discussed. Tables 3.7 through 3.9 show additional mechanical properties of metals, polymers, and ceramics employed in the design of biomaterials in the orthopedic field. These materials will be further discussed in the following sections.

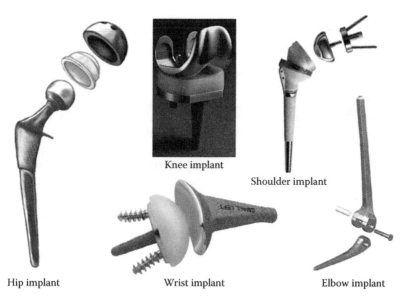

Figure 3.1 Examples of TJR systems. (Modified from AAOS, Total hip replacement—OrthoInfo, n.d., http://orthoinfo.aaos.org/topic.cfm?topic=a00377, accessed February 13, 2016; Knee implant, Robinson, R.P., *J. Arthroplasty*, 20(1 Suppl 1), 2, 2005; Shoulder implant, Berliner, J.L., et al., *J. Shoulder Elbow Surg.*, 24(1), 150, 2015; Total Elbow Arthroplasty, n.d., http://www.houstonmethodist.org/orthopedics/where-does-it-hurt/elbow/artificial-joint-replacement/, accessed February 13, 2016; Wrist Arthroplasty | Bone and Spine, n.d., http://boneandspine.com/wrist-arthroplasty/, accessed February 13, 2016.)

Table 3.7: Mechanical Properties of Metals and Their Alloys Used in TJR and Fracture Management

Material		Tensile Strength σ_{UTS} (MPa)	Young's Modulus E (GPa)
Titanium (Ti) and its alloys	cpTi (pure Ti)	240–550	103
	Ti–6Al–4V	860	110
	Ti–6Al–7Nb	1024	105
	Ti–5Al–2.5Fe	1033	110
	Ti–Zr	900	
	Ti–13Nb–13Zr	1030	79
	Ti–15Mo–5Zr–3Al	882–975	75
		1099–1312	88–113
	Ti–12Mo–6Zr–2Fe	1060–1100	74–85
	Ti–15Mo–2.8Nb–3Al	812	82
		1310	100
	Ti–15Mo–3Nb–0.3O	1020	82
	Ti–35Nb–5Ta–7Zr	590	55
	Ti–35Nb–5Ta–7Zr–0.4O	1010	66
	Ti–0/20Zr–0/20Sn–4/8Nb–2/4Ta + (Pd,N,O)	750–1200	a
CoCr alloys	Co–Cr–Mo	655–1896	210–253
Stainless steel	Stainless steel 316 L	586–1351	190

Sources: Ratner, B.D. et al., *Biomaterials Science—An Introduction to Materials in Medicine*, 3rd edn., Elsevier, Waltham, MA, 2013; Williams, D., *Essential Biomaterials Science*, 1st edn., Cambridge University Press, Cambridge, U.K., 2014; Agrawal, C.M. et al., *Introduction to Biomaterials: Basic Theory with Engineering Applications*, 1st edn., Cambridge University Press, Cambridge, U.K., 2014; Patel, N.R. and Gohil, P.P., *Int. J. Emerg. Technol. Adv. Eng.*, 2, 91, 2012.
[a] Data not available.

Table 3.8: Mechanical Properties of Polymers Used in Total TJR and Fracture Management Systems

Material	Tensile Strength σ_{UTS} (MPa)	Young's Modulus E (GPa)
Polyethylene (PE)	35	0.88
High-density polyethylene (HDPE)	40	1.8
UHMWPE	21	1
Cross-linked UHMWPE	39 ± 3	—[a]
Vitamin-E–blended UHMWPE	43 ± 2	—[a]

Sources: Katti, K.S., *Colloids Surf. B Biointerfaces*, 39, 133, 2004; Bracco, P. and Oral, E., *Clin. Orthop. Relat. Res.*, 469, 2286, 2011.
[a] Data not available.

3.2.1 Total Joint Replacement Systems

Total joint replacement is one of the most efficacious and cost-effective procedures in surgery [28,30,31]. This technique aims to remove a part of arthritic or damaged joint and replace it with biomedical devices that are able to reestablish the movement of a normal and healthy joint [32]. Several joints in the human body such as hip, knee, elbow, wrist, and fingers can be replaced by TJRs

Table 3.9: Mechanical Properties of Ceramics Used in Total TJR

Material	Young's Modulus E (GPa)	Tensile Strength σ_{UTS} (MPa)
Zirconia	220	820
Alumina	380	300
Bioglass	75	—[a]
Hydroxyapatite	117	50

Source: Katti, K.S., *Colloids Surf. B Biointerfaces*, 39, 133, 2004.
[a] Data not available.

(Figure 3.1). However, the joints that usually experience earlier failure, due to arthritis or injuries, are the hip and knee articulations.

3.2.1.1 Total Hip Replacements

Hip implants are mostly composed of four parts, an acetabular cup, a liner, a femoral head, and a femoral stem. Each component requires specific mechanical and biological properties, and several material combinations are available in the market [1]. These devices are subjected to cyclic loading as high as 10^6 cycles per year and in physical activities, such as jumping, they can be exposed up to 10 times the body weight [9]. These conditions require hip components to be resistant against wear generation to avoid material degradation *in vivo*. Materials with high toughness and wear resistance are selected to support these conditions. The stem of femoral components and acetabular cups are usually designed with Ti alloys and in less extension with stainless steel and CoCr alloys [1]. Ti alloys are the most used materials because of their good mechanical properties, biocompatibility, and corrosion resistance as summarized in Tables 3.4 and 3.7 [13]. Although Ti alloys are characterized by high toughness, their Young's Modulus is lower than that observed for stainless steel and CoCr alloys, which avoid problems associated with stress shielding [1,9,13]. Therefore, titanium alloys are the most suitable materials to design femoral stems and acetabular cups. The area of these components that interface with bone is usually surface treated with ceramic and composite nanoparticles to assist with osseointegration [38].

The most varied components of hip implants are the femoral head and the liner. Materials used in these applications are selected to enable the joint to move smoothly with minimal friction and wear generation. The combination of different materials on these components produces designs known as metal-on-metal, metal-on-polyethylene, metal-on-ceramic, ceramic-on-polyethylene, and ceramic-on-ceramic.

Metal-on-metal was composed of a CoCr femoral head and a CoCr cup/liner. However, the generation of metal particles *in vivo*, which was observed to result in aseptic loosening, pseudotumors, periprosthetic osteolysis, and metal hypersensitivity [39], was associated with this design.

Metal-on-polyethylene, ceramic-on-polyethylene, and ceramic-on-ceramic are the systems that are most employed currently and can be selected to attend distinct anatomical requirements of each patient. UHMWPE is the first choice of material as the acetabular liner in metal-on-polyethylene or ceramic-on-polyethylene total hip replacement (THR) systems because of its cost, biocompatibility, and good mechanical properties [40–42]. However, when sliding against a metal or ceramic, wear of UHMWPE is generated due to the weaker interactions among its polymeric chains compared to the interactions among atoms in ceramics and metals [40].

Table 3.10: Mechanical Tests Performed in Hip Implant Components

Testing Type	Description
Hip stem fatigue testing (ISO 7206-6)	The body environment is simulated using predetermined geometry where hip implants are fixed. An axial load of 5340 N is applied and the implants are expected to endure a minimum of 10^6 cycles without failing.
Range of motion and impingement range of motion and impingement (ASTM F2585)	The movement, fatigue, deformation, and wear of the femoral head in the acetabular cup under dynamic impingement condition are tested. A constant joint load of 600 N is applied with a maximum test frequency of 3 Hz up to 10^6 cycles or until the material fails.
Fretting corrosion of hip device connections	The head–neck junction is exposed to saline solution or protein-containing solution, and cyclic load is applied for a minimum of 10^6 cycles. After testing, metal ion is quantified in the fluid and surfaces are microscopically inspected for fretting and corrosion features.

Sources: ASTM F2582—14, Standard test method for impingement of acetabular prostheses, i, 2013, pp. 1–5, doi:10.1520/F2582-08.2; ASTM F1875—98, Standard practice for fretting corrosion testing of modular implant interfaces: Hip femoral headbore and cone taper interface (Reapproved 2009), 2009, pp. 1–6, doi:10.1520/F1875-98R09.2; ISO7206-6:2013, Implants for surgery—Partial and total hip joint prostheses—Part 6: Endurance properties testing and performance requirements of neck region of stemmed femoral components, 2013, pp. 1–15.

Ceramic materials in orthopedics have received a series of improvements in the last few years, which resulted in articulations with minimum wear and little periprosthetic osteolysis [43]. Recent studies show that when wear is generated, ceramic particles induce less inflammatory reactions compared to the particles of UHMWPE [43].

The tests for hip implant components require the ability of prostheses to be able to withstand normal static and dynamic loading under physiological conditions without presenting fracture, plastic deformation, or fatigue fracture. The main tests are listed in Table 3.10.

3.2.1.2 Total Knee Replacements

Total knee replacements (TKRs) are as popular as THRs are and have a similar combination of materials, except for the fact that fewer designs are available for this application. High strength, wear, and corrosion resistance are some of the required properties for the materials used to machine the components of TKR. This device comprises a stemmed tibial plate, a PE-bearing surface, and a contoured metal implant fit around the end of the femur [9,13,30]. Problems associated with TKR are mostly related to the aseptic component loosening and wear of the tibial plate, both related to the generation of micro and macro particles at the bearing articulating surfaces. The stemmed tibial plate is usually designed with titanium alloy due to its good biocompatibility and mechanical properties and, as the stem and acetabular cup components of hip implants, is coated with CaP ceramics to induce osseointegration [9,13,30]. This component is responsible for minimizing the deformation of the PE-bearing insert that is placed between the stemmed plate and femoral component under loads. The contoured metal around the femurs is predominantly made of CoCr, or oxidized zirconium (OxZr), due to the superior wear properties of these materials [47,48]. The selection of CoCr or OxZr has not shown different results in terms of wear generation. Although OxZr femoral components have shown benefits associated with patients with hypersensitivity toward nickel, cobalt, chromium, titanium, and vanadium, recent reports have

Table 3.11: Mechanical Tests Performed in Knee Implant Components

Testing Type	Description
Wear testing (ISO 14243-1, ISO 14243-3, ASTM F2025)	Complex motion of the human body is simulated with the knee prosthesis components using loads from 168 to 2600 N and flexion (0°–58°) applied until crack formation or 10^6 cycles is completed. The internal/external and anterior/posterior motions are based on healthy patient kinematics (about ±5° of internal/external rotation, and 9 mm of total anterior–posterior displacement). Wear between the tibial insert and the femoral component is evaluated according to ASTM F2025.
Fatigue test (ASTM F1800 and ISO 14879)	A half of the tibial plate is fixed using a clamp fixture and the other unsupported half of the plate is subjected to a constant amplitude load up to 10^6 cycles or until failure. Femoral components are evaluated for fatigue properties. Different cementing configurations and flexion angles may be selected to evaluate fatigue.

Sources: ASTM F2025-06, Standard practice for gravimetric measurement of polymeric components for wear assessment (Reapproved 2012), 2012, pp. 1–6, doi:10.1520/F2025-06R12.2.50; ISO 14243-1:2009, Implants for surgery—Wear of total knee-joint prostheses—Part 1: Loading and displacement parameters for wear-testing machines with load control and corresponding environmental conditions for test, 2013, pp. 1–14; ASTM F1800-07, Standard test method for cyclic fatigue testing of metal tibial tray components of total knee joint replacements, 2012, pp. 1–6, doi:10.1520/F1800-12.2; ISO 14879-1:2000, Implants for surgery—Total knee-joint prostheses—Part 1: Determination of endurance properties of knee tibial trays, 2000, pp. 1–7.

shown that this surface can be prone to gouging, which increases the roughness on the bearing surface accelerating PE wear [47,48]. Due to the higher Young's modulus of these materials, which can cause stress shielding, these components are cemented to the bone using commercially available bone cements.

To ensure selected biomaterials meet the mechanical, tribological, and biological requirements, ISO and ASTM have defined standards summarized in Table 3.11.

3.2.2 Fracture Management Systems

Biomaterials are also employed for internal fracture fixation and stabilization as well as for bone reconstruction and replacement [9,30]. Two classes of materials have been proposed for these purposes. The first and most employed class comprises metals such as stainless steel, titanium, and CoCr alloys used in fracture fixation devices. The second class of devices is represented by materials such as polymers and ceramics used as bone void fillers. As mentioned earlier, these materials may have the ability to degrade with time allowing the patient's tissue to regenerate. Moreover, bioactive molecules including growth factors and antibiotics can be employed to improve the healing process of hard and soft tissues.

3.2.2.1 Fracture Fixation

Metal plates, screws, and pins made of biocompatible and corrosion-resistant materials (e.g., titanium and stainless steel) are usually used to repair bone fractures. The success of the procedure will depend on the mechanical environment emergent from the bone–material integration under physiological loads. After the surgical procedure, the healing process and the deformation of the augmented site after surgery under load will be dependent on the characteristics of the fixation device (plate), materials (Ti or stainless steel), geometry (breadth, thickness, and length), and the implanted configurations of screws. The mechanical properties of both Ti and stainless steel are summarized in Tables 3.4 and 3.7. Titanium is superior due to its good mechanical properties, corrosion resistance, and biocompatibility [6].

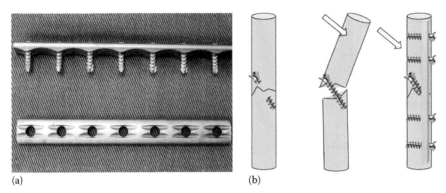

(a)   (b)

Figure 3.2   Example of (a) point-contact fixator and (b) fixation procedure. (Modified from Uhthoff, H.K. et al., *J. Orthop. Sci.*, 11, 118, 2006; Principles of internal fixation—Skeletal trauma, n.d., http://z0mbie.host.sk/Principles-of-Internal-Fixation.html, accessed February 13, 2016.)

On the other hand, devices designed with stainless steel have demonstrated problems with degradation *in vivo* due to corrosion [6]. This particularly happens in junctions between plates and screws, where there is fretting and where the environment can become acidic accelerating surface degradation.

The advantage of these devices in comparison to external limb casts is that they allow early and pain-free movement of the injured limb. Furthermore, these materials avoid the consequences of long-lasting immobilization, such as bone resorption and fracture disease due to immediate and absolute immobilization achieved through interfragmentary compression [9,30]. Examples of materials and applications are illustrated in Figure 3.2a and b. Figure 3.2a shows a point-contact fixator that is locked by screws to achieve angular stability and to prevent bone from being pulled towards the plate. Figure 3.2b demonstrates schematically how the fixation procedure is performed.

Fixation devices can be placed in direct contact with bone; however, in patients that have low bone density or disease conditions such as osteoporosis, cement is used for bone augmentation. Some novel approaches to fixation involve using biodegradable magnesium alloys [19,55]. The use of a biodegradable material eliminates the need for secondary surgery and also lessens the stress shielding effects of the metals. The main issue using magnesium is its quick initial resorption upon implantation.

3.2.2.1.1 Bone Void Fillers

The two main types of bone fillers employed are polymers and ceramics. Due to their distinct properties, these materials are used with different goals. Polymer-based fillers are usually nonresorbable and are used when strength is the primary property needed to stabilize the void or fracture. Ceramic fillers are bioactive and hence provide the opportunity for bone regeneration into the fracture. This regenerative ability comes at a cost of material strength, since these materials are brittle in nature and susceptible to hydrophilic degradation. Both types of fillers are known as bone cements, and their mechanical properties are summarized in Table 3.12.

The tests required to evaluate the mechanical properties of bone cements are described by the ASTM and are described in Table 3.13 [58–60].

3.2.2.1.2 Polymer Cements

The most common polymer cement used is PMMA due to its biocompatibility and mechanical strength (Table 3.12). The cement is provided as two

Table 3.12: Mechanical Properties of Common Bone Cements

	Bone Cement	Tensile Strength (MPa)	Shear Strength (MPa)	Bending Strength (MPa)	Compressive Strength (MPa)	Young's Modulus (GPa)
Polymer	PMMA	35.3	42.2	48.2 ± 0.5	83.0 ± 6.5	5.3 ± 0.5
Ceramic (CaP)	Hydroxyapatite	3.5 ± 0.9	9.8 ± 2.6	—[a]	75.0 ± 4.2	13.5 ± 0.8
	Brushite	1.3 ± 0.3	2.9 ± 0.4	—[a]	10.7 ± 2.0	7.9 ± 0.3
Composite	1.65:1 (PMMA:hydroxyapatite)	—[a]	—[a]	64.0 ± 3.1	103.5 ± 3.0	1.76 ± 0.0
	1.65:1 (PMMA:brushite)	—[a]	—[a]	63.6 ± 2.2	75.6 ± 3.0	1.75 ± 0.3

Sources: Charrière, E. et al., *Biomaterials*, 22, 2937, 2001; Aghyarian, S. et al., *J. Biomater. Appl.*, 29, 688, 2014.

[a] Data not available.

Table 3.13: Mechanical Tests Performed in Bone Cements

Testing Type	Description
Compressive tests (ASTM D695—15)	Bone cement samples need to be prepared as cylinders with a 6 mm diameter and 12 mm height. Specimens need to have their center aligned with the center of the plunger to ensure equal load distribution. The samples are compressed at a controlled rate of 20 mm/min until failure with force and displacement recorded simultaneously.
Flexural 3-point bend test (ASTM D790—15)	Bone cement samples are molded into rectangular bars with approximate dimensions of 70 mm length, 13 mm width, and 4 mm height. The test specimen is placed flat on two supports and deflected medially at a rate of 2.0 mm/min. As deflection displacement increases, the sample experiences compressive forces on the upper surface and tensile forces on the lower surface of the rectangular bar. The test is run until a set strain limit of 5.0% or failure, whichever occurs first.
Fatigue test (ASTM F2118—10)	Bone cement is molded into dumbbell specimens and placed in a uniaxial fatigue setting. An environmental chamber is used to perform the test in phosphate buffer saline solution maintained at 37°C. The cement is then subjected to cyclic tensile and compressive loading at various stress levels at 15 Hz for 5,000,000 cycles. This establishes a number of cycles to failure for every stress level tested.

constituents, a powder and liquid, which harden upon mixing due to a chemical polymerization reaction. The powder phase contains PMMA beads used to speed up the reaction and increase swelling, dibenzoyl peroxide (BPO) that initiates the polymerization, and, if need be, a radiopacifier such as barium sulfate to provide contrast. Contrast is needed for injectable cements as the surgeon must be able to visualize the cement. The liquid phase is composed of methyl methacrylate monomer that will polymerize to form long chains of PMMA, and N,N-dimethyl-p-toluidine that activates the BPO catalyst. The surgeon will mix the powder and liquid and then apply it to the fracture area. The cement will then set, providing stabilization to the site. After setting, PMMA can have compressive strengths ranging from 70 to 103 MPa [57], depending on molecular weight. Furthermore, PMMA is bioinert and will not degrade over time. This can be a hindrance in some cases where bone regeneration needs to be aided and not completely replaced, leading to the next type of cements.

3.2.2.1.3 Ceramic Cements

CaP is the main group of ceramic cement used. CaPs are bioactive and have been shown to enhance bone growth [61]. Some forms can also resorb in the body, allowing eventual bone regrowth into the defect. These cements do not undergo polymerization but harden as CaP crystals in aqueous solution begin to precipitate. This provides a swelling effect, which aids in packing the cement in the defect. The setting process is much slower compared to that of polymer cements, and it results in a highly porous structure, 40%–60%, which can contribute to the CaP cements' brittle nature [62]. Hydroxyapatite ($Ca_5((PO_4)_3(OH))$) and brushite (dicalcium phosphate dehydrate ($CaHPO_4 \cdot 2H_2O$)) are common CaPs utilized as cements. Hydroxyapatite is chemically similar to the inorganic constituent of bone, and due to its high calcium-to-phosphate ratio, 5:3, it is not resorbable. On the other hand, brushite can be resorbed, as it has a calcium-to-phosphate ratio of 1:1. CaP cements are desirable due to their bioactivity; however, their major limitation is their mechanical stability. Bone void fillers are mainly used to stabilize fractured structures, requiring mechanical support provided by the material introduced.

A vertebroplasty study, augmenting vertebral compression fractures, showed recollapse of the defect site after augmenting with CaP cement [63]. Resorbability is desirable; however, it introduces a dynamic of mechanical strength degradation over time with cements that are already brittle. Another limitation is the cements' poor handling properties, making injection very difficult [64].

3.2.2.1.4 Composite Cements
The different properties of the various bone void fillers provide both advantages and limitations. A composite cement approach looks to marry the essential mechanical properties of polymers with the bioactivity of ceramics. Despite this, simple powder–liquid mixing, adding CaP to the polymer powder phase, has been shown to degrade the properties of the composite material. This could be attributed to incomplete mixing leading to aggregation of CaP in the polymer matrix. The formed heterogeneous structure is susceptible to failure due to these stress-rising areas. In response, novel cement systems are being developed that can accommodate PMMA and CaP without deteriorating side effects [57,65,66].

3.3 APPLICATIONS OF BIOMATERIALS IN DENTAL IMPLANTS
Dental implants are biomaterials developed to provide structural and functional restorations for teeth. Based on anatomical tissue locations, there are four major types of dental implant systems, namely, (1) intramucosal, (2) subperiosteal, (3) transosteal, and (4) endosseous. Intramucosal, subperiosteal, and transosteal implants are used when there is limited bone availability [67]. Nowadays, these implant systems are not routinely employed. Currently, the endosseous root-form implant system is the most widely utilized replacement for partial and complete edentulous patients. In the United States, 3 million people already have dental implants placed and around 500,000 patients receive implants every year [68]. One million implants are placed every year worldwide [69]. Biomechanics of this particular system will be discussed in this topic.

3.3.1 Endosseous Dental Implants
An endosteal implant is similar to the structure of a natural tooth. The prosthetic crown is supported by an implant submerged inside the bone, similar to that of a tooth with the crown being supported by the root (Figure 3.3). Hence, they are also known as root-form endosseous implants [70]. While in a natural tooth, the crown is the continuation of the root; different components are combined together to create a dental implant. The implant and the crown setup is connected together internally either by a cement-retained or a screw-retained abutment [71] as shown in Figure 3.3b.

3.3.2 Biomaterials for Dental Implants
The modern-day dental implants were essentially developed from an accidental discovery in 1952 by a Swedish doctor named Per-Ingvar Branemark [74]. Branemark's medical research found that if titanium and bone were left undisturbed in contact with one another, bone would grow right up against the titanium surface. This phenomenon, osseointegration, led to the eventual development of the endosseous dental implants known today. Commercially pure titanium (CpTi) (Grade 4) became the material of choice. The Ti6Al4V alloy is also used [75]. Recently, titanium–ZrO_2 alloys were introduced in the dental implants' market [9]. Thereafter, studies were performed to promote successful osseointegration and faster bone healing with various types of surface treatments imparted on bulk metals. A summary of a few leading dental implant–related industries, their choice of bulk metal, and surface modifications is listed in Table 3.14.

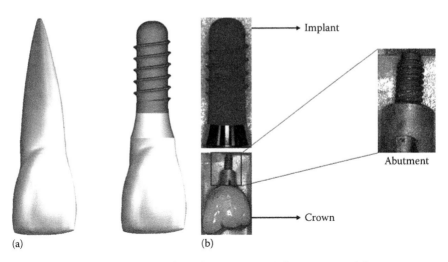

(a)   (b)

Figure 3.3 Structure of Dental Implant System: (a) Comparison of the structure of a dental implant with a natural tooth; (b) Components of a Dental Implant System. ([a]: Modified from What is a dental implant? Dental implants are a permanent solution to replace missing teeth and are an alternative to dentures, n.d., http://www.dentalimplants.uchc.edu/about/index.html, accessed February 13.)

Table 3.14: Bulk Metals and Their Surface Treatments for Commercially Available Dental Implants

Manufacturer	Material	Surface Modification
Straumann	Cp Ti, Ti/Zr alloys	Sandblasted, large-grit, acid-etched (SLA) surface
Zimmer-Biomet 3i	Cp Ti, Tantalum	Micro-textured (MTX) surface, MTX-treated and hydroxyapatite-coated surface Discrete crystal deposition (DCD) surface
Nobel Biocare	Cp Ti	Ti Unite surface, fabricated by anodic oxidation
Dentsply	Cp Ti	OsseoSpeed surface. Grit blasted with TiO_2 and acid etched

Sources: OsseoSpeed™—More bone more rapidly (n.d.), http://www.dentsplyimplants.com/~/media/M3Media/DENTSPLYIMPLANTS/ScientificDocumentation/57148 Scientific review OsseoSpeed.ashx?filetype=.pdf (accessed February 13, 2016); Basic information on the surgican procedures (n.d.), http://ifu.straumann.com/content/dam/internet/straumann_ifu/brochures/152.754_low.pdf (accessed February 13, 2016); Trabecular Metal™ dental implants: Overview of design and developmental research (n.d.), http://www.zimmerdental.com/tm/pdf/tm_TMWhitePaper2096.pdf (accessed February 13, 2016); Lausmaa, J., *Appl. Osseointegration Res.*, 1, 5, 2000.

An implant is connected to an abutment or post, a modular piece which protrudes through the gum and supports the crown or prosthetic tooth. From a prosthetic point of view, these systems allow for more flexibility in the range of restorations, which can be accomplished. Currently, Cp Ti and ZrO_2 are the predominant materials for fabrication of abutments [80]. Studies have shown that metallic abutments have a tendency to present a visible bluish tinge through the soft or gum tissue. This led to the growing popularity of ZrO_2 that overcame this problem and provided a better esthetic appearance [81]. Metal-related allergy in patients placed with dental implants has been reported in a few clinical studies [16]. Recent experiments have shown the ability of ZrO_2 to osseointegrate [82] and have encouraged researchers to explore the possibilities of fabricating ceramic-based implants [83].

3.3.2.1 Biomechanics of Dental Implants

Traditionally, the surgical placement of dental implants involves two stages. In the first stage, the implant will remain submerged inside the bone flap (known as "osteotomy"). This is known as the healing period during which an implant is allowed to achieve osseointegration in an undisturbed environment. This period is usually around 3–4 months and 5–6 months for implants placed in maxilla and mandible, respectively [84]. After achieving successful osseointegration, prosthetic components such as the abutment and crown will be attached to the implant. A fully restored implant will be exposed to mastication/occlusal loading or functional loading. Henceforth, these two stages will give rise to two different aspects of biomechanics in dental implants: (1) biomechanics of peri-implant tissues during the healing phase; (2) biomechanics of implants and peri-implant bone after exposure to occlusal loading or after osseointegration.

3.3.2.2 Biomechanics of Peri-Implant Tissues during Healing Phase

The concept of passive healing in implant dentistry hypothesizes that maxillar and mandibular bones transfer some loads from the bone to the endosteal implant during this healing period [85]. The implant achieves osseintegration and stability during this phase. There are two kinds of stability: (1) primary stability and (2) secondary stability.

Primary stability is the absence of mobility of an implant immediately after surgical placement and is a phenomenon of purely biomechanical nature, related mostly to bone quality at the implant site [86]. It reduces the risk of micromovements preventing fibrous tissue formation and most importantly allows bone apposition on the implant surface [87]. Implant design, surgical protocol, congruence between bone and implant, and friction phenomena at the bone–implant interface are influential factors for primary stability [88]. Friction at the bone–implant interface defines the torque required for surgical insertion of an implant. Clinical results indicate that an insertion torque of about 40 N cm increases the success rate of implants [89]. There are also studies that reported 14% of failed implants when inserted with torques in the range of 25–35 N cm [90].

Secondary stability is achieved by bone remodeling and subsequent bone apposition bridging old bone and implant surface after 3–6 months of the healing period [87]. In this process of healing, bone quality at osteotomy site, surface microtopography, and chemistry of implants play a vital role [91]. Surface properties such as hydrophilicity, wettability, chemistry, and topography of implants are proposed to result in bone filling simultaneously from bone-to-implant and implant-to-bone directions [92]. The extension of bone from implant toward healing bone is said to be 30% faster than bone healing toward the surface [93]. Recent developments in surgical instrumentation have made it possible to measure implant stability to confirm whether osseointegration is successful enough to allow functional loading [29]. These are electroacoustic techniques that include (1) Periotest [94], (2) resonance frequency analysis [95], and (3) quantitative ultrasound methods [96].

3.3.2.3 Biomechanics of Implant and Peri-Implant Bone after Osseointegration

According to Branemark and Skalak, osseointegration offers a stable and apparently immobile support of prosthesis under functional loads without pain, inflammation, (or) loosening [97]. Both the implant-supported prosthesis and the peri-implant bone will be subjected to mechanical stress under functional

Table 3.15: Comparison between Teeth and Implant

Functionality	Teeth	Implant
Fulcrum to lateral forces	Apical 1/3 of root	Crestal bone
Lateral movement	56–108 μm	10–50 μm
Apical movement	25–100 μm	3–5 μm

Source: Chen, Y.Y. et al., *J. Dent. Sci.*, 3, 65, 2008.

loading. It is important to understand that the anatomy of an implant *in vivo* is largely different from that of a tooth. A natural tooth is surrounded by periodontal ligament (PDL), which serves as a shock absorber and helps in force distribution. The absence of ligament tissues in an osseointegrated dental implant restricts movement, and mechanical stress is concentrated on the peri-implant crestal bone. In a natural tooth, neurophysiological innervations of PDL detect changes in occlusal forces encountered. The absence of these receptors in peri-implant tissues will increase sensitivity even to small applied forces [98]. A comparison of mobility and force distribution between dental implants is given in Table 3.15.

An average individual approximately performs 1×10^6 chewing cycles per year [99]. After a year of functional loading, peri-implant bone undergoes remodeling and a bone loss of up to 2 mm is considered normal [100]. Previous studies elucidated that excessive stress can cause microfractures on the implant and within bone leading to eventual bone loss [101]. Occlusal overloading has been associated with peri-implantitis and implant failure [102]. Overloading can sometimes result due to parafunction such as bruxism, which will lead to excessive involuntary clenching of teeth [103]. Frost proposed a theory that states that there is a minimal effective strain above which adaptive response would take place, while below it, bone would remain stable [104]. However, a maximum load limit that an implant can withstand without jeopardizing the surrounding bone tissues remains unanswered [105].

Branemark and Skalak further warned that there should be no progressive relative motion between the implant and the surrounding tissues [97]. This suggests that relative motion or micromotion between the implant and peri-implant bone is inevitable [87]. The threshold level for micromotion of an implant in the bone–implant interface has been reported in the literature to be approximately 150 μm [106]. Abnormal overloading exceeding threshold relative motion can lead to microdamage of bone due to fatigue. This microdamage to interfacial bone is hypothesized to result in increased bone remodeling eventually causing implant loosening [107]. However, related mechanisms of these displacements are still not completely understood and studies about several biomechanical characters are inconclusive [108]. Cyclic occlusal loads imposed on dental implants can also cause micromotion of the modular parts of an implant [47]. Implants typically have two modular parts: (1) a modular connection between the screw and abutment, and (2) a second modular area between the abutment and crown. These modular junctions have been widely investigated for their tendency to generate microgaps at the mating region, which are reported to range from 30 to 200 μm [109]. Modular junction micromotion has been reported to cause microorganism leakage [110], crestal bone changes [111], mechanical instability [51], screw fracture, and in some cases implant failure [112].

3.3.2.4 Tools for Evaluating Biomechanical Performance

There are a number of tools that are currently utilized when working with and characterizing dental implant systems. Computer-aided design (CAD) tools are commonly employed to design, adapt upon, and evaluate dental implant and dental implant systems in a computational world prior to manufacturing physical parts to save time and money. The clinical viability and long-term success of dental implants (as with many other implant systems) depends in part on the mechanical integrity of the implant material in the environment in which it is placed. These implants must be designed to withstand mastication forces for up to 10 years. Henceforth, *in vitro* bench-top tests are also performed to simulate the *in vivo* biomechanical environment and assess the performance of implants. As discussed earlier, criteria/standards for successful performance are provided by international or federal agencies such as the ISO and the U.S. Food and Drug Administration (FDA). ISO 14801 deals with the biomechanical test specifications for dental implants. These methods commonly consist of static compressive force testing, dynamic fatigue testing, and tensile retentive strength testing.

3.3.2.4.1 Finite Element Analysis

Finite element analysis (FEA) is the most widely utilized computational tool employed in the design, characterization, and testing of dental implants. FEA is a method for reliably predicting the effects of forces, vibrations, heat, and fluid flow on computerized models of physical objects using advanced mathematics [105]. These tools allow for the definition of each set or portion of the object, and the definition of how the objects will be loaded, displaced, heated, etc. The mechanical properties of the materials must all be known and inputted into the software. Then certain questions can be evaluated. As an example, if a researcher or designer wanted to know what would happen to a dental implant after a specific force was applied (e.g., another tooth biting down on it), the user could import the implant into the FEA software, define and apply loads, and quickly solve the stresses and strains in the resulting equilibrium.

3.3.2.4.2 Computational Fluid Dynamics

Because of the predominantly mechanical nature of dental implant systems, fluid dynamics has not traditionally been employed in their design, development, and characterization. However, recent studies by Wadhwani et al. have begun investigating the utility of these computational fluid dynamic (CFD) tools in understanding more about the mechanics of the dental cement components within abutment-crown cement seating procedures. Residual dental cement has recently been added to the American Academy of Periodontology's list of peri-implant disease risk factors [113–115]. Because these cements are designed to retain the dental crown prosthesis onto the abutment—but not to be interacting with physiologic tissues—it is important to ensure that the abutment designs and cementation techniques consider the cement flow patterns, volume of cement, and cement rheology. These tools are currently being used to better understand how incorporating ventilation into the abutment design can improve dental cement flow throughout the implant system, improve the retention of the crown, and reduce the risk of residual cement leakage [113]. These investigations typically involve building an abutment design using CAD software and incorporating that design into a CFD environment. Here, fluid properties can be entered into the software and fluid starting locations can be defined prior to starting the experiment. Once all the starting variables are accounted for, simulations can be run to determine the fluid mechanics in different designs and systems. The outputs can aid in the improvement or redesign of certain aspects of the system.

3.3.2.4.3 Static Compressive Force Testing

Static compressive force analysis is important in the characterization of dental implants for two reasons. First, it is important to understand the maximum force that can be applied to a dental implant to ensure that a single bite onto the implant system will not cause the implant to fail (crack, break, or deform). Human maximum bite forces range between 200 and 400 N [116]. Second, the dynamic fatigue testing (described in the following) relies on this characterization as a starting point for deciding initial load values to begin dynamic testing. Testing procedures and requirements are detailed in Table 3.16.

Table 3.16: Mechanical Tests Performed on Dental Implant Systems

Testing Type	Description
Static compressive force testing	Implant's long axis makes a 30° angle with the loading direction of the testing machine. Additionally, the implant should be supported 3 mm below the anticipated crestal bone level to simulate at least 3 mm of crestal bone loss [120]. Test implants are tested using quasi-static loading until fracture or plastic deformation occurs using a crosshead speed of approximately 1 mm/min [123,121,118]. Force (N)–displacement (mm) curves are then generated to determine the yield of the implant systems under load (F). For comparison of fracture resistance with other commercially available systems, often the bending moment (M) is calculated in N cm according to the formula $M = 0.5 \times F \times L$ (ISO Norm 14801) with F being the load (N) and L being the vertical distance from the simulated bone level to the center of the load (cm) [121,118]. Test should be conducted in at least triplicate to ensure statistical reliability in assessment.
Dynamic fatigue testing	Implant's long axis makes a 30° angle with the loading direction of the testing machine. The implant should be also supported 3 mm below the anticipated crestal bone level to simulate at least 3 mm of crestal bone loss [120]. Current ISO standards limit testing in wet conditions to a rate of 2 Hz until failure or 2 million cycles. Dry conditions are limited to 15 Hz until failure or 5 million cycles. If the test implant includes materials which are subject to corrosion fatigue, or include polymeric components, wet testing should be selected (testing in water, normal saline, or physiologic medium). All other systems should be tested in air (dry). The primary implication of testing in dry versus wet conditions is time. Testing at 2 Hz will last until 12 days, while testing at 15 Hz will last until 4 days. Typically, testing begins at 80% of the static failure load with respect to the particular device system. Then the load is iteratively decreased until the endurance limit is reached. Tests are conducted in triplicate over 4 or more loading scenarios.
Tensile retentive strength testing	This characterization often utilizes cast metal crown copings fabricated specifically for the retentive analysis on a test abutment design. The copings will typically have been fabricated with a loop at the occlusal end to allow for the attachment of a jig for the pull-off procedure in the testing system [119]. Dental cement (varies per testing parameters selected for) is loaded into the crowns and subsequently seated to its respective abutment and held with finger pressure until set [114]. The prefabricated cast metal crowns are pulled out under tensile load with a mechanical testing system at a crosshead speed of 1 mm/min until debonding of the crowns or fracture of the abutment or crown occurs [122].

Sources: Wadhwani, C. et al., *Int. J. Oral Maxillofac. Implants*, 26, 1241; Lee, C.K. et al., *Dent. Mater.*, 25, 1419, 2009; ISO 14801:2007, Dentistry—Implants—Dynamic fatigue test for endosseous dental implants, pp. 1–9, 2007; Mayta-Tovalino, F. et al., *J. Dent. Implant.*, 5, 25, 2015; Guidance for industry and FDA Staff Class II special controls guidance document: Root-form endosseous dental implants and endosseous dental implant abutments, Test, pp. 1–20, 2004; Stawarczyk, B. et al., *J. Prosthet. Dent.*, 107, 94, 2012; Gehrke, S.A., *Clin. Implant Dent. Relat. Res.*, 790, 2013.

3.3.2.4.4 Dynamic Fatigue Testing

Dynamic fatigue testing is the primary testing necessary for a dental implant development. It deals with the implant endurance or lifetime. The testing is designed to test how an implant will behave during cyclic loading and unloading over long periods of time. The ISO recommendations (ISO 14801) were originally designed for single endosteal (transmucosal) dental implants testing under "worst case" scenarios [117,118]. Testing procedures and requirements are detailed in Table 3.16.

3.3.2.4.5 Tensile Retentive Strength Testing

Tensile retentive strength analysis of dental implants is concerned with the characterization of the retentive strength between the dental crown (prosthesis) and abutment (modular connection between the crown and implant). Crowns are often cemented atop the abutments, so the crown fabrication (luting/relief space between the crown and abutment) as well as the abutment design can contribute to the retentive strength of these systems [119].

This chapter presented an overall overview of biomaterials used in implantable devices discussing the evolution of classes of materials, their properties, and current applications in the field. A number of other biomaterials are in use in clinical practice, and this chapter limited the discussion on applications in joints and dental devices.

REFERENCES

1. B.D. Ratner, A.S. Hoffman, F.J. Schoen, J.E. Lemons, *Biomaterials Science—An Introduction to Materials in Medicine*, 3rd edn., Elsevier, Waltham, MA, 2013.

2. B.D. Ratner, S.J. Bryant, Biomaterials: Where we have been and where we are going, *Annu. Rev. Biomed. Eng.* 6 (2004) 41–75.

3. D. Williams, *Essential Biomaterials Science*, 1st edn., Cambridge University Press, Cambridge, U.K., 2014.

4. N. Huebsch, D.J. Mooney, Inspiration and application in the evolution of biomaterials, *Nature* 462 (2009) 426–432.

5. R. Langer, D.A. Tirrell, Designing materials for biology and medicine, *Nature* 428 (2004) 487–492.

6. C.M. Agrawal, J.L. Ong, M.R. Appleford, G. Mani, *Introduction to Biomaterials: Basic Theory with Engineering Applications*, 1st edn., Cambridge University Press, Cambridge, U.K., 2014.

7. N.R. Patel, P.P. Gohil, A review on biomaterials: Scope, applications & human anatomy significance, *Int. J. Emerg. Technol. Adv. Eng.* 2 (2012) 91–101.

8. D.F. Williams, On the nature of biomaterials, *Biomaterials* 30 (2009) 5897–5909.

9. K.S. Katti, Biomaterials in total joint replacement, *Colloids Surf. B Biointerfaces* 39 (2004) 133–142.

10. J. Black, G. Hastings, *Handbook of Biomaterials Properties*, 1st edn., Springer, Berlin, Germany, 1998.

11. T. Hanawa, Research and development of metals for medical devices based on clinical needs, *Sci. Technol. Adv. Mater.* 13 (2012) 064102.

12. J. Mouhyi, D.M. Dohan Ehrenfest, T. Albrektsson, The peri-implantitis: Implant surfaces, microstructure, and physicochemical aspects, *Clin. Implant Dent. Relat. Res.* 14 (2012) 170–183.

13. M.B. Nasab, M.R. Hassan, B. Bin Sahari, Metallic biomaterials of knee and hip—A review, *Trends Biomater. Artif. Organs* 24 (2010) 69–82.

14. D. Scheinert, S. Scheinert, J. Sax, C. Piorkowski, S. Bräunlich, M. Ulrich et al., Prevalence and clinical impact of stent fractures after femoropopliteal stenting, *J. Am. Coll. Cardiol.* 45 (2005) 312–315.

15. D.C. Rodrigues, R.M. Urban, J.J. Jacobs, J.L. Gilbert, In vivo severe corrosion and hydrogen embrittlement of retrieved modular body titanium alloy hip-implants, *J. Biomed. Mater. Res. B Appl. Biomater.* 88 (2009) 206–219.

16. D.C. Rodrigues, P. Valderrama, T. Wilson, K. Palmer, A. Thomas, S. Sridhar et al., Titanium corrosion mechanisms in the oral environment: A retrieval study, *Materials (Basel)* 6 (2013) 5258–5274.

17. O.M. Posada, R.J. Tate, M.H. Grant, Effects of CoCr metal wear debris generated from metal-on-metal hip implants and Co ions on human monocyte-like U937 cells, *Toxicol. In Vitro* 29 (2015) 271–280.

18. A.K. Low, G.S. Matharu, S.J. Ostlere, D.W. Murray, H.G. Pandit, How should we follow-up asymptomatic metal-on-metal hip resurfacing patients? A prospective longitudinal cohort study, *J. Arthroplasty* 31 (2015) 146–151.

19. K.F. Farraro, K.E. Kim, S.L.-Y. Woo, J.R. Flowers, M.B. McCullough, Revolutionizing orthopaedic biomaterials: The potential of biodegradable and bioresorbable magnesium-based materials for functional tissue engineering, *J. Biomech.* 47 (2014) 1979–1986.

20. T. Jaeblon, Polymethylmethacrylate: Properties and contemporary uses in orthopaedics, *J. Am. Acad. Orthop. Surg.* 18 (2010) 297–305.

21. S. Li, A.H. Burstein, Ultra-high molecular weight polyethylene. The material and its use in total joint implants, *J. Bone Jt. Surg.* 76 (1994) 1080–1090.

22. S.M. Kurtz, *The UHMWPE Handbook*, 1st edn., Elsevier, San Diego, CA, 2004.

23. P. Bracco, E. Oral, Vitamin E-stabilized UHMWPE for total joint implants: A review, *Clin. Orthop. Relat. Res.* 469 (2011) 2286–2293.

24. L.S. Nair, C.T. Laurencin, Biodegradable polymers as biomaterials, *Prog. Polym. Sci.* 32 (2007) 762–798.

25. T. V. Thamaraiselvi, S. Rajeswari, Biological evaluation of bioceramic materials—A review, *Trends Biomater. Artif. Organs* 18 (2004) 9–17. http://medind.nic.in/taa/t04/i1/taat04i1p9.pdf.

26. F. Baino, G. Novajra, C. Vitale-Brovarone, Bioceramics and scaffolds: A winning combination for tissue engineering, *Front. Bioeng. Biotechnol.* 3 (2015) 202.

27. W.-S. Choy, K.J. Kim, S.K. Lee, K.W. Bae, Y.S. Hwang, C.K. Park, Ceramic-on-ceramic total hip arthroplasty: Minimum of six-year follow-up study, *Clin. Orthop. Surg.* 5 (2013) 174–179.

28. W. Wang, Y. Ouyang, C.K. Poh, Orthopaedic implant technology: Biomaterials from past to future, *Ann. Acad. Med. Singapore* 40 (2011) 237–243.

29. W. Zhai, H. Lu, L. Chen, X. Lin, Y. Huang, K. Dai et al., Silicate bioceramics induce angiogenesis during bone regeneration, *Acta Biomater.* 8 (2012) 341–349.

30. S. Santavirta, Y.T. Konttinen, R. Lappalainen, A. Anttila, S.B. Goodman, M. Lind et al., Materials in total joint replacement, *Curr. Orthop.* 12 (1998) 51–57.

31. P.-A. Vendittoli, N. Duval, D.J. Stitson, B. Mâsse, Vertical acetabular positioning with an inclinometer in total hip arthroplasty, *J. Arthroplasty* 17 (2002) 936–941.

32. J.M. Martell, J.J. Verner, S.J. Incavo, Clinical performance of a highly cross-linked polyethylene at two years in total hip arthroplasty: A randomized prospective trial, *J. Arthroplasty* 18 (2003) 55–59.

33. AAOS, Total hip replacement—OrthoInfo (n.d.). http://orthoinfo.aaos.org/topic.cfm?topic=a00377 (accessed February 13, 2016).

34. R.P. Robinson, The early innovators of today's resurfacing condylar knees, *J. Arthroplasty* 20(1 Suppl 1) (Jan 2005) 2–26.

35. J.L. Berliner, A. Regalado-Magdos, C.B. Ma, B.T. Feeley, Biomechanics of reverse total shoulder arthroplasty, *J Shoulder Elbow Surg.* 24(1) (Jan 2015) 150–160.

36. Total Elbow Arthroplasty (n.d.). http://www.houstonmethodist.org/orthopedics/where-does-it-hurt/elbow/artificial-joint-replacement/ (accessed February 13, 2016).

37. Wrist Arthroplasty | Bone and Spine (n.d.). http://boneandspine.com/wrist-arthroplasty/ (accessed February 13, 2016).

38. X. Liu, P. Chu, C. Ding, Surface modification of titanium, titanium alloys, and related materials for biomedical applications, *Mater. Sci. Eng. R Rep.* 47 (2004) 49–121.

39. M. Burbano, R. Russell, M. Huo, R. Welch, D. Roy, D.C. Rodrigues, Surface characterization of retrieved metal-on-metal total hip implants from patients with adverse reaction to metal debris, *Materials (Basel)* 7 (2014) 1866–1879.

40. E.M. Brach del Prever, A. Bistolfi, P. Bracco, L. Costa, UHMWPE for arthroplasty: Past or future?, *J. Orthop. Traumatol.* 10 (2009) 1–8.

41. O.K. Muratoglu, C.R. Bragdon, D.O. O'Connor, M. Jasty, W.H. Harris, A novel method of cross-linking ultra-high-molecular-weight polyethylene to improve wear, reduce oxidation, and retain mechanical properties, *J. Arthroplasty* 16 (2001) 149–160.

42. E. Oral, S.D. Christensen, A.S. Malhi, K.K. Wannomae, O.K. Muratoglu, Wear resistance and mechanical properties of highly cross-linked, ultra-high-molecular weight polyethylene doped with vitamin E, *J. Arthroplasty* 21 (2006) 580–591.

43. B.-J. Kang, Y.-C. Ha, D.-W. Ham, S.-C. Hwang, Y.-K. Lee, K.-H. Koo, Third-generation alumina-on-alumina total hip arthroplasty: 14 to 16-year follow-up study, *J. Arthroplasty* 30 (2015) 411–415.

44. ASTM F2582—14, Standard test method for impingement of acetabular prostheses, i (2013). pp. 1–5. doi:10.1520/F2582-08.2.

45. ASTM F1875—98, Standard practice for fretting corrosion testing of modular implant interfaces: Hip femoral head-bore and cone taper interface (Reapproved 2009) (2009). pp. 1–6. doi:10.1520/F1875-98R09.2.

46. ISO7206-6:2013, Implants for surgery—Partial and total hip joint prostheses—Part 6: Endurance properties testing and performance requirements of neck region of stemmed femoral components (2013). pp. 1–15.

47. T.C. Gascoyne, M.G. Teeter, L.E. Guenther, C.D. Burnell, E.R. Bohm, D.R. Naudie, In vivo wear performance of cobalt-chromium versus oxidized zirconium femoral total knee replacements, *J. Arthroplasty* 31(1) (Jan 2016) 137–141.

48. J.-M. Brandt, L. Guenther, S. O'Brien, A. Vecherya, T.R. Turgeon, E.R. Bohm, Performance assessment of femoral knee components made from cobalt–chromium alloy and oxidized zirconium, *Knee* 20 (2013) 388–396.

49. ASTM F2025-06, Standard practice for gravimetric measurement of polymeric components for wear assessment (Reapproved 2012) (2012). pp. 1–6. doi:10.1520/F2025-06R12.2.

50. ISO 14243-1:2009, Implants for surgery—Wear of total knee-joint prostheses—Part 1: Loading and displacement parameters for wear-testing machines with load control and corresponding environmental conditions for test (2009). http://www.iso.org/iso/iso_catalogue/catalogue_tc/catalogue_detail.htm?csnumber=44262 (accessed February 13, 2016).

51. ASTM F1800-07, Standard test method for cyclic fatigue testing of metal tibial tray components of total knee joint replacements (2012). pp. 1–6. doi:10.1520/F1800-12.2.

52. ISO 14879-1:2000, Implants for surgery—Total knee-joint prostheses—Part 1: Determination of endurance properties of knee tibial trays (2000).

53. H.K. Uhthoff, P. Poitras, D.S. Backman, Internal plate fixation of fractures: Short history and recent developments, *J. Orthop. Sci.* 11 (2006) 118–126.

54. Principles of internal fixation—Skeletal Trauma (n.d.). http://z0mbie.host.
sk/Principles-of-Internal-Fixation.html (accessed February 13, 2016).

55. A. Weizbauer, M. Kieke, M.I. Rahim, G.L. Angrisani, E. Willbold,
J. Diekmann et al., Magnesium-containing layered double hydroxides as
orthopaedic implant coating materials—An *in vitro* and *in vivo* study, *J.
Biomed. Mater. Res. Part B Appl. Biomater.* 104(3) (Apr 2016) 525–531.

56. E. Charrière, S. Terrazzoni, C. Pittet, P. Mordasini, M. Dutoit, J. Lemaître
et al., Mechanical characterization of brushite and hydroxyapatite cements,
Biomaterials 22 (2001) 2937–2945.

57. S. Aghyarian, L.C. Rodriguez, J. Chari, E. Bentley, V. Kosmopoulos, I.H. Lieberman
et al., Characterization of a new composite PMMA-HA/Brushite bone cement
for spinal augmentation, *J. Biomater. Appl.* 29 (2014) 688–698.

58. ASTM D965—15, Compressive properties of rigid plastics (2015). pp. 1–8.
doi:10.1520/D0695-15.2.

59. ASTM D790—15, Standard test methods for flexural properties of unre-
inforced and reinforced plastics and electrical insulating materials (2015).
pp. 1–12. doi:10.1520/D0790-15.2.

60. ASTM F2118—10, Standard test method for constant amplitude of
force controlled fatigue testing of acrylic bone cement materials, ASTM
International (2010). doi:10.1520/F2118–10, doi:10.1520/F2118-03R09.2.

61. S.V. Dorozhkin, M. Epple, Biological and medical significance of calcium
phosphates, *Angew. Chem. Int. Ed. Engl.* 41 (2002) 3130–3146.

62. S. Becker, M. Ogon, *Balloon Kyphoplasty*, 1st edn., Springer Vienna, Vienna,
Austria, 2008.

63. A. Piazzolla, G. De Giorgi, G. Solarino, Vertebral body recollapse without
trauma after kyphoplasty with calcium phosphate cement, *Musculoskelet.
Surg.* 95 (2011) 141–145.

64. G. Baroud, M. Bohner, P. Heini, T. Steffen, Injection biomechanics of bone
cements used in vertebroplasty, *Biomed. Mater. Eng.* 14 (2004) 487–504. http://
www.ncbi.nlm.nih.gov/pubmed/15472396 (accessed February 13, 2016).

65. D.C. Rodrigues, N.R. Ordway, C.R.-J. Ma, A.H. Fayyazi, J.M. Hasenwinkel,
An ex vivo exothermal and mechanical evaluation of two-solution bone
cements in vertebroplasty, *Spine J.* 11 (2011) 432–439.

66. L. Rodriguez, J. Chari, S. Aghyarian, I. Gindri, V. Kosmopoulos,
D. Rodrigues, Preparation and characterization of injectable brushite
filled-poly (methyl methacrylate) bone cement, *Materials (Basel)* 7 (2014)
6779–6795.

67. B. Yeshwante, S. Patil, N. Baig, S. Gaikwad, A. Swami, M. Doiphode, Dental
implants—Classification, success and failure—An Overview, *IOSR J. Dent.
Med. Sci. Ver. II* 14 (2015) 2279–2861.

68. D. Duraccio, F. Mussano, M.G. Faga, Biomaterials for dental implants: Current and future trends, *J. Mater. Sci.* 50 (2015) 4779–4812.

69. Dental implants facts and figures (n.d.). http://www.aaid.com/about/ Press_Room/Dental_Implants_FAQ.html (accessed February 4, 2016).

70. A. Gupta, M. Dhanraj, G. Sivagami, Status of surface treatment in endosseous implant: A literary overview, *Indian J. Dent. Res.* 21 (2010) 433–438.

71. R. Shadid, N. Sadaqa, A comparison between screw- and cement-retained implant prostheses. A literature review, *J. Oral Implantol.* 38 (2012) 298–307.

72. What is a dental implant? Dental implants are a permanent solution to replace missing teeth and are an alternative to dentures (n.d.). http://www. dentalimplants.uchc.edu/about/index.html (accessed February 13, 2016).

73. Dental implants—Prosper family dentistry (n.d.). http://www. prosperfamilydentistry.com/dental-implants/ (accessed February 13, 2016).

74. P.I. Branemark, Osseointegration and its experimental background, *J. Prosthet. Dent.* 50 (1983) 399–410.

75. J. Gottlow, M. Dard, F. Kjellson, M. Obrecht, L. Sennerby, Evaluation of a new titanium-zirconium dental implant: A biomechanical and histological comparative study in the mini pig, *Clin. Implant Dent. Relat. Res.* 14 (2012) 538–545.

76. OsseoSpeed™—More bone more rapidly (n.d.). http://www.dentsplyimplants. com/~/media/M3 Media/DENTSPLY IMPLANTS/Scientific Documentation/57148 Scientific review OsseoSpeed.ashx?filetype=.pdf (accessed February 13, 2016).

77. Basic information on the surgican procedures (n.d.). http://ifu.straumann. com/content/dam/internet/straumann_ifu/brochures/152.754_low.pdf (accessed February 13, 2016).

78. Trabecular Metal™ dental implants: Overview of design and developmental research (n.d.). http://www.zimmerdental.com/tm/pdf/tm_ TMWhitePaper2096.pdf (accessed February 13, 2016).

79. J. Lausmaa, Properties of a new porous oxide surface on titanium implants, *Appl. Osseointegration Res.* 1 (2000) 5–8.

80. J.K.W. Foong, R.B. Judge, J.E. Palamara, M.V. Swain, Fracture resistance of titanium and zirconia abutments: An in vitro study, *J. Prosthet. Dent.* 109 (2013) 304–312.

81. A. Sicilia, S. Cuesta, G. Coma, I. Arregui, C. Guisasola, E. Ruiz et al., Titanium allergy in dental implant patients: A clinical study on 1500 consecutive patients, *Clin. Oral Implants Res.* 19 (2008) 823–835.

82. R. Depprich, H. Zipprich, M. Ommerborn, E. Mahn, L. Lammers, J. Handschel et al., Osseointegration of zirconia implants: An SEM observation of the bone-implant interface, *Head Face Med.* 4 (2008) 25.

83. M. Andreiotelli, H.J. Wenz, R.J. Kohal, Are ceramic implants a viable alternative to titanium implants? A systematic literature review, *Clin. Oral Implants Res.* 20 (2009) 32–47.

84. M.G. Newman, H. Takei, P.R. Klokkevold, F.A. Carranza, *Clinical Periodontology*, 10th edn., Elsevier, Saint Louis, MO, 2015.

85. J.E. Lemons, Biomaterials, biomechanics, tissue healing, and immediate-function dental implants, *J. Oral Implantol.* 30 (2004) 318–324.

86. N.F. Oliscovicz, A.C. Shimano, É. Marcantonio Junior, C.P. Lepri, A.C. Dos Reis, Analysis of primary stability of dental implants inserted in different substrates using the pullout test and insertion torque, *Int. J. Dent.* 2013 (2013) 1–5.

87. G. Haïat, H.-L. Wang, J. Brunski, Effects of biomechanical properties of the bone-implant interface on dental implant stability: From in silico approaches to the patient's mouth, *Annu. Rev. Biomed. Eng.* 16 (2013) 187–213.

88. S. Hansson, M. Norton, The relation between surface roughness and interfacial shear strength for bone-anchored implants. A mathematical model, *J. Biomech.* 32 (1999) 829–836.

89. S. Anil, P.S. Anand, H. Alghamdi, J.A. Jansen, Dental implant surface enhancement and osseointegration, in: I. Turkyilmaz (Ed.), *Implant Dentistry—A Rapidly Evolving Practice*, 1st edn., In Tech, Rijeka, Croatia, 2011, pp. 83–108.

90. G. Cannizzaro, M. Leone, V. Ferri, P. Viola, G. Federico, M. Esposito, Immediate loading of single implants inserted flapless with medium or high insertion torque: A 6-month follow-up of a split-mouth randomised controlled trial, *Eur. J. Oral Implant.* 5 (2012) 333–342. http://ejoi.quintessenz.de/index.php?doc=abstract&abstractID=28937/ (accessed February 13, 2016).

91. Y. Shibata, Y. Tanimoto, A review of improved fixation methods for dental implants. Part I: Surface optimization for rapid osseointegration, *J. Prosthodont. Res.* 59 (2015) 20–33.

92. U. Gross, Biomechanically optimized surface profiles by coupled bone development and resorption at hydroxyapatite surfaces, *Soc. Biomater. Trans.* 13 (1990) 83.

93. D.A. Puleo, A. Nanci, Understanding and controlling the bone-implant interface, *Biomaterials* 20 (1999) 2311–2321.

94. S.O. Koutayas, T. Vagkopoulou, S. Pelekanos, P. Koidis, J.R. Strub, Zirconia in dentistry: Part 2. Evidence-based clinical breakthrough, *Eur. J. Esthet. Dent. Winter* 4 (2009) 348–380.

95. N. Meredith, D. Alleyne, P. Cawley, Quantitative determination of stability of the implant-tissue interface using resonance frequency analysis, *Clin. Oral Implants Res.* 7 (1996) 261–267.

96. V. Mathieu, F. Anagnostou, E. Soffer, G. Haïat, Ultrasonic evaluation of dental implant biomechanical stability: An in vitro study, *Ultrasound Med. Biol.* 37 (2011) 262–270.

97. R. Brånemark, P.-I. Brånemark, B. Rydevik, R.R. Myers, Osseointegration in skeletal reconstruction and rehabilitation, *J. Rehabil. Res. Dev.* 38 (2001) 1–4.

98. Y.Y. Chen, C.L. Kuan, Y.B. Wang, Implant occlusion: Biomechanical considerations for implant-supported prostheses, *J. Dent. Sci.* 3 (2008) 65–74.

99. S.R. Allum, R.A. Tomlinson, R. Joshi, The impact of loads on standard diameter, small diameter and mini implants: A comparative laboratory study, *Clin. Oral Implants Res.* 19 (2008) 553–559.

100. F.S. Rosen, P. Clem, D. Cochran, Peri-implant mucositis and peri-implantitis: A current understanding of their diagnosis and clinical implications, *J. Periodontol.* 84 (2013) 436–443.

101. M. Esposito, J.M. Hirsch, U. Lekholm, P. Thomsen, Biological factors contributing to failures of osseointegrated oral implants. (II) Etiopathogenesis, *Eur. J. Oral Sci.* 106 (1998) 721–764.

102. J.-H. Fu, Y.-T. Hsu, H.-L. Wang, Identifying occlusal overload and how to deal with it to avoid marginal bone loss around implants, *Eur. J. Oral Implantol.* 5 Suppl. (2012) S91–S103.

103. K. Nishigawa, E. Bando, M. Nakano, Study of bite force during sleep associated bruxism, *J. Oral Rehabil.* 28 (2001) 485–491.

104. R.L. Duncan, C.H. Turner, Mechanotransduction and the functional response of bone to mechanical strain, *Calcif. Tissue Int.* 57 (1995) 344–358.

105. A. Sertgöz, S. Güvener, Finite element analysis of the effect of cantilever and implant length on stress distribution in an implant-supported fixed prosthesis, *J. Prosthet. Dent.* 76 (1996) 165–169.

106. W. Winter, D. Klein, M. Karl, Micromotion of dental implants: Basic mechanical considerations, *J. Med. Eng.* 2013 (2013) 1–9.

107. D.B. Burr, R.B. Martin, M.B. Schaffler, E.L. Radin, Bone remodeling in response to in vivo fatigue microdamage, *J. Biomech.* 18 (1985) 189–200.

108. S. Saidin, M.R. Abdul Kadir, E. Sulaiman, N.H. Abu Kasim, Effects of different implant-abutment connections on micromotion and stress distribution: Prediction of microgap formation, *J. Dent.* 40 (2012) 467–474.

109. C. do Nascimento, R.F. de Albuquerque Jr., Bacterial leakage along the implant-abutment interface, in: I. Turkyilmaz (Ed.), *Implant Dentistry—Most Promising Discipline of Dentistry*, 1st edn., In Tech, Rijeka, Croatia, 2011, pp. 324–347. http://www.intechopen.com/books/implant-dentistry-the-most-promising-discipline-of-dentistry/bacterial-leakage-along-the-implant-abutment-interface\nInTech (accessed February 8, 2017).

110. M. Quirynen, D. van Steenberghe, Bacterial colonization of the internal part of two-stage implants. An in vivo study, *Clin. Oral Implants Res.* 4 (1993) 158–161.

111. C.M. Stanford, R.A. Brand, Toward an understanding of implant occlusion and strain adaptive bone modeling and remodeling, *J. Prosthet. Dent.* 81 (1999) 553–561.

112. A.R.C. Duarte, J.P.S. Neto, J.C.M. Souza, W.C. Bonachela, Detorque evaluation of dental abutment screws after immersion in a fluoridated artificial saliva solution, *J. Prosthodont.* 22 (2013) 275–281.

113. C. Wadhwani, *Cementation in Dental Implantology: An Evidence-Based Guide*, 1st edn., Springer, Berlin, Germany, 2015.

114. C. Wadhwani, A. Piñeyro, T. Hess, H. Zhang, K.-H. Chung, Effect of implant abutment modification on the extrusion of excess cement at the crown-abutment margin for cement-retained implant restorations, *Int. J. Oral Maxillofac. Implants* 26 (2011 Nov–Dec), 1241–1246.

115. T. Linkevicius, E. Vindasiute, A. Puisys, V. Peciuliene, The influence of margin location on the amount of undetected cement excess after delivery of cement-retained implant restorations, *Clin. Oral Implants Res.* 22 (2011) 1379–1384.

116. P. Takaki, M. Vieira, S. Bommarito, Maximum bite force analysis in different age groups, *Int. Arch. Otorhinolaryngol.* 18 (2014) 272–276.

117. C.K. Lee, M. Karl, J.R. Kelly, Evaluation of test protocol variables for dental implant fatigue research, *Dent. Mater.* 25 (2009) 1419–1425.

118. ISO 14801:2007, Dentistry—Implants—Dynamic fatigue test for endosseous dental implants (2007). pp. 1–9.

119. F. Mayta-Tovalino, J. Rosas-Díaz, V. Ccahuana-Vasquez, Removal force of cast copings to abutments with three luting agents, *J. Dent. Implant.* 5 (2015) 25.

120. Guidance for industry and FDA Staff Class II special controls guidance document: Root-form endosseous dental implants and endosseous dental implant abutments, Test (2004) pp. 1–20.

121. F. Massoglia, A. Catalan, A. Martinez, M. Flores, Bending moments and failure of titanium and zirconia abutments with internal connections: In vitro study, *J. Dent. Heal. Oral Disord. Ther. Bend.* 4 (2016) 1–7.

122. B. Stawarczyk, T. Basler, A. Ender, M. Roos, M. Özcan, C. Hämmerle, Effect of surface conditioning with airborne-particle abrasion on the tensile strength of polymeric CAD/CAM crowns luted with self-adhesive and conventional resin cements, *J. Prosthet. Dent.* 107 (2012) 94–101.

123. S.A. Gehrke, Importance of crown height ratios in dental implants on the fracture strength of different connection designs: An in vitro study, *Clin. Implant Dent. Relat. Res.* (2013) 790–797. doi:10.1111/cid.12165.

4 Biomaterials in Cancer Research

From Basic Understanding to Applications

Edna George and Shamik Sen

CONTENTS

4.1 INTRODUCTION

Cancer arises from the uncontrolled proliferation of cells leading to the formation of clinically detectable solid tumors. While benign tumors stay in their original locations, high mortality in cancers is attributed to metastasis, that is, the spreading of cancer cells to distant tissues eventually leading to their destruction. Behavior of cancer cells is dictated by factors in the tumor microenvironment; these include the extracellular matrix (ECM), blood vessels, and noncancerous cells including immune cells and fibroblasts. The ECM exhibits dynamic remodeling in response to pathological condition. Tumor development is accompanied with the stiffening of the surrounding matrix (Egeblad et al. 2010; Provenzano et al. 2006), which results in alterations of nearly 1500 genes as reported in human mammary epithelial cells (Alcaraz et al. 2008). Proliferation of cancer cells in a confined environment results in a microenvironment that is oxygen deprived (i.e., hypoxic); upregulation of hypoxia-inducible factor (HIF-1) in turn activates the vascular endothelial growth factor (VEGF) (Liao and Johnson 2007) leading to sprouting of new blood vessels (angiogenesis) (Ferrara 2002). Additionally, the tumor microenvironment consists of normal fibroblasts that are activated to form cancer-associated fibroblasts (CAFs) (Kalluri and Zeisberg 2006). These CAFs play a crucial role in the development of tumor, angiogenesis as well as metastasis (Elenbaas and Weinberg 2001; Orimo et al. 2005; Vered et al. 2010). The cancer cells slowly breach the basement membrane, invade the blood vessels, and relocate to distant locations in the body (Figure 4.1). In recent years, in addition to the enormous research carried out to understand the genetic and molecular basis of cancer, given the role of ECM alterations in driving cancer progression (Lu et al. 2012), and the close cross talk between cancer cells and stromal cells (Pollard 2004), targeting the tumor microenvironment has been the focus of many evolving therapies (Engels et al. 2012).

To delineate the cellular and molecular mechanism of cancer, several studies have been carried out by culturing the monolayer of cancer cells. These 2D culture studies are important in understanding many genetic predispositions to cancer that involve the regulation of protooncogenes (e.g., RAS, ERB, TRK) as well as tumor suppressor genes (e.g., RB, p53, APC) (Weinberg 1994). Interestingly, one of the best known tumor suppressor genes of today, p53, was initially considered as a protooncogene. Later, several gene mapping studies carried out with primary specimens and cell lines of colorectal carcinoma established p53 as a tumor suppressor gene (Baker et al. 1989; Miyashita et al. 1994). While several studies have been carried out to understand cancer progression independent of

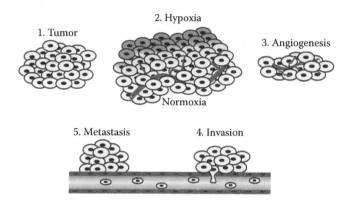

Figure 4.1 Cancer progression. Cancer arises due to uncontrolled prolif-eration of cells (step 1) leading to a hypoxic environment that is deprived of oxygen (step 2). This results in the sprouting of new blood vessels known as angiogenesis (step 3). The tumor cells breach the basement membrane, invade the vasculature (step 4), and metastasize to distant locations within the body (step 5).

its microenvironment, the role of microenvironment-specific cells including fibro-blasts and macrophages is underexplored. To address the physical and chemical interactions between cancer cells and stromal cells, co-culturing is required. For example, enhanced invasiveness of breast carcinoma cells in the presence of bone marrow fibroblasts is associated with enhanced MMP secretion by fibroblasts (Saad et al. 2000). CAFs were also shown to be capable of inducing epithelial-to-mesenchymal transition in breast cancer cells (Gao et al. 2010).

Although 2D culture techniques and co-culturing of different cell lines have helped to understand the molecular and genetic mechanisms of cancer develop-ment, given the importance of the physicochemical attributes of the tumor microen-vironment, researchers have tried to recapitulate one or more of these cues *in vitro*. For example, in contrast to adding hypoxic factors to induce hypoxia in 2D culture (Choi et al. 2011), hypoxia is naturally set up in 3D tumor spheroids (Menrad et al. 2010). 3D microenvironment can also be recapitulated in xenograft murine models where human cancer cells are grown within immunocompromised mice (Karnoub et al. 2007). On the downside, monitoring these immunocompromised mice for several weeks over the course of treatment becomes difficult. Additionally, though these murine models give a better understanding of the tumor microenvironment, an uncertainty prevails in translating these findings to human application. Several clinical trials have proved the drug to be inefficient for treating human cancers, although clinical trials in murine models have been successful (Mak et al. 2014). Moreover, in tumor spheroids and xenograft murine models, control of architec-tural, mechanical, and biological factors is not possible. Thus, mimicking the tumor microenvironment together with the complexity of the tumor dynamics has been the rate limiting step in understanding cancer progression.

Tumor spheroids and xenograft mouse models are widely used as models for studying and understanding cancer. But given the skill required to pre-pare uniform spheroids and the high cost associated with murine xenograft models, researchers have turned their attention to biomaterials to recapitu-late the tumor microenvironment for studying various aspects of tumor pro-gression. Chemical properties like pH (Deng et al. 2011), surface chemistry (Thevenot et al. 2008; Yang et al. 2012), wetting properties (Gerber et al. 2006;

Kaplan et al. 2016), mechanical properties like stiffness (Soman et al. 2012), viscosity (Dhiman et al. 2004; Zawaneh et al. 2010), material architecture like topology, porosity, and particle size (Mitragotri and Lahann 2009) will determine the biological versatility of the material. Biomaterials can also serve as an important tool to develop new strategies for imaging deep-set tumors as well as to design treatment modalities for treating cancer. Here, we will discuss about the different biomaterials used in cancer research.

4.2 BIOMATERIALS FOR STUDYING CANCER INVASION

Matrigel—the gelatinous protein mixture secreted by Engelbreth-Holm-Swarm mouse sarcoma cells that polymerize to form a gel at 37°—represents one of the earliest biomaterials used for studying cancer invasion (Albini et al. 1987; Kleinman et al. 1986). While matrigels have been used for cancer invasion studies and can even recapitulate the formation of vascular networks with endothelial cells, batch-to-batch differences in the composition of ECM proteins and growth factors along with the low content of collagen I (Gelain et al. 2006; Sodek et al. 2008) have led researchers to search for other biomaterials that provide better tunability.

In the zest to better understand tumor dynamics, researchers have focused on developing scaffolds with control of cell adhesive moieties, biocompatibility, and degradability—the factors that are altered during cancer progression and influence invasiveness of cancer cells (Beck et al. 2013; Xu et al. 2012). These polymers have been used in various forms ranging from nanoparticles to thick hydrogels for an array of studies including cell migration (Chaw et al. 2007), ECM degradation, secretion of ECM proteins, angiogenesis (Hielscher and Gerecht 2012), drug delivery (Lin and Metters 2006), imaging, and therapeutics (Lei et al. 2011). Given the ECM stiffening (Figure 4.2) observed in multiple epithelial cancers (Egeblad et al. 2010; Kass et al. 2007), researchers have focused their attention on developing hydrogel systems where stiffness can be tuned over a wide range. Polyacrylamide (PA) (Levental et al. 2009), polydimethylsiloxane (Liu et al. 2009), and polyethylene glycol diacrylate (PEGDA) (Soman et al. 2012) hydrogels are some of the widely used synthetic scaffolds used for studying cancer invasion. Ease of fabrication, ease of imaging, ability to tune stiffness over a wide range (100 Pa–200 kPa), and ease of functionalization with any ECM protein have made PA hydrogels one of the most widely used synthetic polymers used for

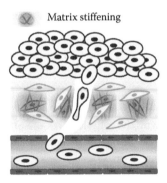
Matrix stiffening

Figure 4.2 ECM remodeling and cancer invasion. Many epithelial cancers are characterized by increased deposition of fibrillar proteins including collagen I, and its cross-linking, leading to increase in the matrix stiffness. ECM stiffening leads to cancer invasion through enhanced integrin signaling.

cancer studies (Nemir and West 2010). Using PA gels, researchers have shown that increased stiffness is associated with higher cell proliferation (Wang et al. 2000), increased spreading (Collin et al. 2006), increased motility, and increased cell–ECM tractions (Yeung et al. 2005) across multiple different cancer cells. In a seminal study, using PA gels of stiffness encompassing that of normal and malignant mammary stroma, Valerie Weaver and coworkers demonstrated that normal breast epithelial cells cultured in soft collagen ECMs (<150 Pa) form normal acinar epithelial structures, whereas those cultured in rigid ECMs (>1 kPa) form dysplastic tissue architectures with uncontrolled cell division, reminiscent of a growing tumor (Le et al. 2009). These observations were subsequently reproduced in a mouse model wherein lysyl oxidase–mediated collagen cross-linking stiffened the ECM, promoted focal adhesion formation, enhanced PI3 kinase (PI3K) activity, and induced cellular invasion (Rubashkin et al. 2014).

While synthetic scaffolds have contributed significantly to our knowledge of cancer invasion, most of these studies are on 2D. For studies focused on understanding 3D cancer cell invasion, researchers have made use of natural polymers including silk (Kwon et al. 2010), alginate (Kievit et al. 2010), hyaluronic acid (Gurski et al. 2009; Xu et al. 2012), collagen (Huang et al. 2013; Provenzano et al. 2006), modified gelatin (Huang et al. 2008; Tamura et al. 2015), and fibrin (Brown et al. 1993; Liu et al. 2012). Unlike the stiffness dependence of cell motility in 2D, increase in stiffness in 3D scaffolds suppresses cancer invasion due to reduction in pore size (Yang et al. 2010). Collagen gels represent one of the most widely used hydrogel systems for studying cancer invasion (Schor et al. 1982; Xu and Burg 2007). Using 3D collagen matrices, seminal work by Wolf and Weiss groups have illustrated the role of confinement in inhibiting cell invasion (Wolf et al. 2013), the critical role of MT1-MMP in mediating invasion through ECM degradation (Sabeh et al. 2004), and nuclear deformability as the rate limiting factor in cell invasion (Davidson et al. 2014).

Apart from increased invasiveness of cancer cells, tumor angiogenesis is critical for cancer metastasis. In a study focused on understanding the angiogenic capacity of squamous cell carcinoma cells in 3D arginine–glycine–aspartic acid (RGD) peptide-bound alginate gels, upon integrin engagement, cells were found to secrete high levels of IL-8 and VEGF, factors which influence the spatial control of tumor vascularization (Fischbach et al. 2009). In another elegant study, fluorinated ethylene propylene tubing was linked to a collagen gel to recreate a neo-vessel by co-culturing MDA-MD-231 breast cancer cells with telomerase immortalized microvascular endothelial cells under tumor-relevant hydrodynamic shear flow conditions. In this study tumor vascular model highlighted the importance of the hydrodynamic tumor microenvironment in regulating angiogenic activity, which was established by the increase in tumor-expressed angiogenic genes under 3D dynamic conditions (Buchanan et al. 2014b). Studies have also shown angiogenesis modified mechanosensitivity and drug sensitivity using a composite hydrogel system of PEGDA and methacrylated gelatin (Wu et al. 2015). Recently, a 3D *in vitro* model was developed using methacrylated gelatin to fabricate microwells that were used to grow glioblastoma spheroids on endothelial monolayers. This cost-effective setup established cell communication between the 3D spheroid and 2D monolayer leading to angiogenesis (Nguyen et al. 2016).

After escaping from the primary tumor, cancer cells enter into the blood vessels where they are subjected to blood flow–induced shear stresses. Previous studies have shown the effect of shear stress on endothelial cell reorganization and angiogenesis (Dewey et al. 1981; Franco et al. 2016; Kohn et al. 2015). Shear stress has also been shown to influence tumor angiogenesis and drug resistance (Santoro et al. 2015). Interestingly, co-culture of breast tumor cells and endothelial cells in the microfluidic tumor vascular model designed to mimic wall shear stresses felt by endothelial cells and tumor cells (Buchanan et al. 2014a) revealed an inverse relation between wall shear stress and endothelial permeability. In another study, upon exposure to shear stresses in a perfusion bioreactor, malignant bone cancer cells cultured on electrospun poly(ε-caprolactone) scaffolds were found to activate insulin-like growth factor-1 (IGF-1) receptor (IGF-1R) pathway. Since these receptors are expressed in Ewing's sarcoma, IGF-1 can be considered as a reliable target for preclinical and early-phase drug development (Fong et al. 2013; Subbiah et al. 2009).

One of the reasons for cancer relapse is attributed to the presence of cancer stem cells (CSCs)—slow cycling cells that possess higher metastatic potential and are unresponsive to widely used chemotherapeutic drugs (Clevers 2011). While CSCs are typically enriched using surface markers enzyme activity (Duan et al. 2013), and mammosphere assays (Saadin and White 2013), maintaining CSCs in their undifferentiated state remains a challenge. Biomaterials have been developed to enrich and expand CSCs (Ordikhani et al. 2015). For example, PEGDA has been useful in the selective enrichment of CSCs owing to the absence of any bioadhesive ligands (Yang et al. 2013). PEGDA hydrogels with stiffness ranging from 200 to 7000 Pa were used to study the optimum stiffness required for the growth and marker expression CSCs. Cancer cells that originated from different tissues of differing stiffness were studied, which concluded that the optimum stiffness for CSCs is dependent on the origin of the cancer cells (Jabbari et al. 2015).

4.3 BIOMATERIALS FOR TUMOR IMAGING

Poor prognosis due to lack of specific symptoms and lack of early treatment has been some of the leading causes of deaths due to cancer. Though late-stage tumors can be physically detected as palpable lumps (e.g., in breast cancer), currently available techniques detect tumors at a later stage and are often unable to detect deep-set tumors. Interdisciplinary research in nanotechnology and biomedical engineering has made great strides in making early detection and treatment possible. Several physiochemical features of cancer can be exploited to illuminate tumor at an early stage, thereby enabling efficient treatment and making complete recovery possible (Yu et al. 2012). Consequently, ongoing research by several research groups is focusing on developing biocompatible nanoparticles for tumor detection given their size-tunable properties, noninvasiveness, choice of materials that are nonantigenic, and the freedom to engineer stealth properties (Li and Huang 2010). The nanoparticles can be utilized with maximum efficiency by exploiting the leaky vasculature, which represents one of the hallmarks of cancer (Danhier et al. 2010). This can help in the passive targeting of tumor sites and gaining access to deep-set tumors (Schleich et al. 2014). In addition to passive targeting, the nanoparticles can be actively targeted by functionalization with ligands specific to surface receptors expressed by cancer cells, and also by changes in pH (Poon et al. 2011). The advantages of using nanoparticles include size-dependent emission spectra as seen in quantum dots (Michalet et al. 2005), chemical modification for multimodal imaging (Figure 4.3) (Perrault et al. 2009), and size-dependent bio-distribution (Yhee et al. 2014).

Major studies carried out in cancer imaging have been associated with gadolinium-doped nanoparticles for excellent contrast properties in magnetic resonance imaging (MRI) (Zhou and Lu 2013). An example is the peptide nanoparticles conjugated with near-infrared (NIR) fluorescence emitting dye for imaging pancreatic cancer (Montet et al. 2006). In this study, the presence of bombesin receptors in normal pancreas and the absence of the same in pancreatic

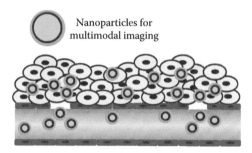

Nanoparticles for multimodal imaging

Figure 4.3 Multimodal tumor imaging using nanoparticles. The difficulty to image deep-set tumors makes it necessary to combine multiple imaging modalities within a single platform. Leaky tumor vasculature allows for accumulation of nanoparticles at the tumor site, thereby enabling its detection.

ductal adenocarcinoma were utilized to design bombesin-bound iron oxide nanoparticles conjugated with Cy5.5 dye with a particle size of 35 nm. These nanoparticles exhibited higher binding affinity to normal pancreatic cells and with better contrast, thereby enhancing tumor visualization. Decreased T2 value was observed in the normal areas of the pancreas while the tumor region showed enhanced T2 values. Similar nanoparticles were designed to target underglycosylated mucin-1 antigen for *in vivo* imaging of the mouse model (Moore 2004). Another material used in cancer imaging is bioresorbable calcium phosphate nanoparticles (CPNPs) that can be doped with a variety of contrast agents (Ashokan et al. 2013). One such example is the use of CPNPs doped with NIR-emitting fluorophore, indocyanine green (ICG) (Altinoğlu et al. 2008). In this study, ICG-CPNPs exhibited 200% higher quantum efficiency compared to the free fluorophore. Twenty-four hours post tail vein injection in a nude mouse, these CPNPs accumulated within breast adenocarcinoma tumors via enhanced retention and permeability (EPR) effect, thereby allowing an imaging depth up to 3 cm.

Given the various modes of imaging, combining the different modes to improve the quality of the image by depth imaging gave rise to multimodal imaging. For example, hydroxyapatite nanocrystals (nHAP) of ~30 nm were engineered to combine MRI, x-ray imaging, and (NIR) fluorescence imaging within a single entity (Ashokan et al. 2010). These nanoparticles were doped with rare earth element Eu^+ and co-doped with Gd^+ to obtain nanocrystals suitable for multimodal imaging. To enhance the efficiency of accumulation, these nanocrystals were grafted with folic acid that specifically bind to cancer cells expressing folate receptors (Ashokan et al. 2010). *In vivo* medical imaging using protein nanoparticles has also gained importance owing to their biological stability and biocompatibility. *Escherichia coli* DNA-binding protein (spherical, 10 nm) and *Thermoplasma acidophilum* proteasomes (cylindrical, 12 × 15 nm) are two such protein nanoparticles that were engineered with RGD insertion, and NIR dye bound to lysine residue. These RGD-presenting protein nanoparticles were taken up by integrin-expressing tumor cells (Ahn et al. 2014). Thus, biomaterials for cancer imaging range from polymeric nanomaterials decorated with ligands to biologically stable protein nanoparticles to image deep-set tumors.

4.4 BIOMATERIALS FOR CANCER THERAPEUTICS

Detrimental and prolonged side effects of chemotherapy paved way for an alternative strategy for treating cancers. Biomaterials play an important role in treating cancers by assisting in administering small molecules, genes, and

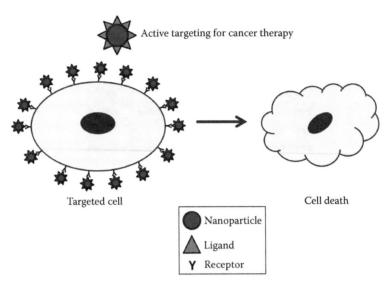

Active targeting for cancer therapy

Targeted cell

Cell death

● Nanoparticle

▲ Ligand

Y Receptor

Figure 4.4 Cancer therapeutics. Functionalization of nanoparticles using a ligand specific to a receptor overexpressed in cancer cells can be used to actively target cancer cells. Selective targeting of cancer cells also reduces the harmful effects of drugs on normal cells.

peptides using nano-formulations, including liposomes, micelles, and dendrimers. Translation of these formulations for clinical applications is expected to reduce drug sensitization at nonspecific sites, localization and release of drugs at the tumor site via active or passive targeting (Torchilin 2010; Yokoyama 2005), retaining of the drugs at target sites in optimal quantities (Jain 1998), higher cellular uptake (Chou et al. 2011; Ding and Ma 2013), and biodegradability (Nair and Laurencin 2007). Many nano-drugs are passively targeted to retain drug circulation for longer periods so that it accumulates at the tumor site owing to the EPR effect (Maeda et al. 2000). On the other hand, drug encapsulated polymeric systems can be actively targeted by conjugating the system with target ligands like folate, galactosamine, epidermal growth factor, and transferrin against receptors that are overexpressed in specific cancers (Figure 4.4).

Liposomes have been of immense interest as drug delivery agents owing to the presence of hydrophilic and hydrophobic compartments and allow for circumventing the problems of insolubility of most anticancer drugs in water (Akbarzadeh et al. 2013; Voinea and Simionescu 2002). In addition to their biocompatibility (owing to their lipid bilayers), liposomes offer the luxury of carrying both water-soluble and water-insoluble drugs within a single entity. Modified strategies have been developed with the aim of delivering drugs to specific tumor sites. A study showed the use of pH-sensitive liposomes for site-specific delivery of doxorubicin with reduced systemic side effects for treating breast cancer. These liposomes were tagged with estrone to target the estrogen receptors of breast cancer. The pH at the target site triggered the pH-sensitive liposome to release the membrane-impermeable molecules to the cytosol (Paliwal et al. 2012). Another study demonstrated the use of thermosensitive liposomes complexed with pH-sensitive liposomes to deliver doxorubicin. The aim of incorporating lysolipid in thermosensitive liposomes was to aid the release of the drug under mild hyperthermia by destabilizing the drug carrier forcing the drug to be released at the target site (Kheirolomoom et al. 2013).

Another group of biomaterials that are used in cancer drug delivery are the dendrimers that have a 3D spherical shape ranging from 1 to 10 nm in diameter with a core of initiator, repeated branching units, terminal functional groups, and void spaces that form rooms for the cargo. Some of the commonly used dendrimers are poly(ethylene oxide) (PEO), polyamidoamines (PAMAM), poly(L-lysine) scaffold dendrimers, polyesters (PGLSA-OH), polypropylimines (PPI), and poly(2,2-bis(hydroxymethyl)propanoic acid) scaffold dendrimers (Bis-MPA) (Jin et al. 2014). Studies using PAMAM dendrimers have shown charge-based bio-distribution in melanoma and human prostate cancer mouse model systems, with the net surface charge of the dendrimers dictating its bio-distribution. Although the trend was similar, higher deposition was observed in lungs, liver, and kidneys compared to that in tumor, heart, pancreas, and spleen; the lowest levels were found in the brain. Additionally, dendrimers with a net positive charge showed higher deposition than dendrimers with a net neutral charge (Nigavekar et al. 2004). In another study, asymmetric dendrimers were created using PEO to form bow-tie dendrimers loaded with doxorubicin to inhibit the progression of doxorubicin-insensitive C-26 tumors. The bow-tie architecture of the dendrimers allows flexibility in varying the length of the PEO chain and the drug-loading sites (Lee et al. 2006). Dendrimers of PAMAM origin have also been used to actively target tumors by conjugating dendrimers using folic acid (Singh et al. 2008). These dendrimers of <5 nm were used to target human KB tumors grown in immunodeficient mice (Kukowska-Latallo et al. 2005).

4.5 CONCLUSION

In this chapter, we have discussed about three classes of biomaterials used in cancer research that span the areas of cancer invasion, cancer imaging, and cancer treatment. The use of biomaterials for mimicking cancer has allowed researchers to probe the individual contributions of matrix remodeling, angiogenesis, and cell migration to cancer invasion. Furthermore, use of nanoparticles in imaging has allowed access to deep-set tumors that would have otherwise remained inaccessible to currently available techniques. Tagging therapeutic effects to the nanoparticles has enabled simultaneous imaging and treatment of cancer. In short, biomaterials have contributed significantly to our knowledge of cancer, its detection and treatment.

REFERENCES

Ahn, K.-Y., H.K. Ko, B.-R. Lee, E.J. Lee, J.-H. Lee, Y. Byun, I.C. Kwon, K. Kim, and J. Lee. 2014. Engineered protein nanoparticles for in vivo tumor detection. *Biomaterials* 35(24): 6422–6429.

Akbarzadeh, A., R. Rezaei-Sadabady, S. Davaran, S.W. Joo, N. Zarghami, Y. Hanifehpour, M. Samiei, M. Kouhi, and K. Nejati-Koshki. 2013. Liposome: Classification, preparation, and applications. *Nanoscale Research Letters* 8(1): 102.

Albini, A., Y. Iwamoto, and H.K. Kleinman. 1987. A rapid in vitro assay for quantitating the invasive potential of tumor cells. *Cancer Research* 47: 3239–3245. http://ovidsp.ovid.com/ovidweb.cgi?T=JS&PAGE=reference&D=emed1b&NEWS=N&AN=1987150155. Accessed March 24, 2016.

Alcaraz, J., R. Xu, H. Mori, C.M. Nelson, R. Mroue, V.A. Spencer, D. Brownfield, D.C. Radisky, C. Bustamante, and M.J. Bissell. 2008. Laminin and biomimetic extracellular elasticity enhance functional differentiation in mammary epithelia. *The EMBO Journal* 27(21): 2829–2838.

Altinoğlu, E.I., T.J. Russin, J.M. Kaiser, B.M. Barth, P.C. Eklund, M. Kester, and J.H. Adair. 2008. Near-infrared emitting fluorophore-doped calcium phosphate nanoparticles for in vivo imaging of human breast cancer. *ACS Nano* 2(10): 2075–2084.

Ashokan, A., G.S. Gowd, V.H. Somasundaram, A. Bhupathi, R. Peethambaran, A.K.K. Unni, S. Palaniswamy, S.V. Nair, and M. Koyakutty. 2013. Multifunctional calcium phosphate nano-contrast agent for combined nuclear, magnetic and near-infrared invivo imaging. *Biomaterials* 34(29): 7143–7157.

Ashokan, A., D. Menon, S. Nair, and M. Koyakutty. 2010. A molecular receptor targeted, hydroxyapatite nanocrystal based multi-modal contrast agent. *Biomaterials* 31(9): 2606–2616.

Baker, S.J., E.R. Fearon, J.M. Nigro, S.R. Hamilton, A.C. Preisinger, J.M. Jessup, P. VanTuinen et al. 1989. Chromosome 17 deletions and p53 gene mutations in colorectal carcinomas. *Science* 244(4901): 217–221. http://eutils.ncbi.nlm.nih.gov/entrez/eutils/elink.fcgi?dbfrom=pubmed&id=2649981&retmode=ref&cmd=prlinks. Accessed March 22, 2016.

Beck, J.N., A. Singh, A.R. Rothenberg, J.H. Elisseeff, and A.J. Ewald. 2013. The independent roles of mechanical, structural and adhesion characteristics of 3D hydrogels on the regulation of cancer invasion and dissemination. *Biomaterials* 34(37): 9486–9495 (Elsevier Ltd).

Brown, L.F., B. Berse, R.W. Jackman, K. Tognazzi, E.J. Manseau, D.R. Senger, and H.F. Dvorak. 1993. Expression of vascular permeability factor (vascular endothelial growth factor) and its receptors in adenocarcinomas of the gastrointestinal tract. *Cancer Research* 53(18): 4727–4735.

Buchanan, C.F., S.S. Verbridge, P.P. Vlachos, and M.N. Rylander. 2014a. Flow shear stress regulates endothelial barrier function and expression of angiogenic factors in a 3D microfluidic tumor vascular model. *Cell Adhesion & Migration* 8: 517–524.

Buchanan, C.F., E.E. Voigt, C.S. Szot, J.W. Freeman, P.P. Vlachos, and M.N. Rylander. 2014b. Three-dimensional microfluidic collagen hydrogels for investigating flow-mediated tumor-endothelial signaling and vascular organization. *Tissue Engineering. Part C, Methods* 20(1): 64–75.

Chaw, K.C., M. Manimaran, E.H. Tay, and S. Swaminathan. 2007. Multi-step microfluidic device for studying cancer metastasis. *Lab on a Chip* 7(8): 1041–1047.

Choi, J.Y., Y.S. Jang, S.Y. Min, and J.Y. Song. 2011. Overexpression of MMP-9 and HIF-1alpha in breast cancer cells under hypoxic conditions. *Journal of Breast Cancer* 14(2): 88–95.

Chou, L.Y.T., K. Ming, and W.C.W. Chan. 2011. Strategies for the intracellular delivery of nanoparticles. *Chemical Society Reviews* 40(1): 233–245.

Clevers, H. 2011. The cancer stem cell: Premises, promises and challenges. *Nature Medicine* 17(3): 313–319.

Collin, O., P. Tracqui, A. Stephanou, Y. Usson, J. Clément-Lacroix, and E. Planus. 2006. Spatiotemporal dynamics of actin-rich adhesion microdomains: Influence of substrate flexibility. *Journal of Cell Science* 119(Pt 9): 1914–1925.

Danhier, F., O. Feron, and V. Préat. 2010. To exploit the tumor microenvironment: Passive and active tumor targeting of nanocarriers for anti-cancer drug delivery. *Journal of Controlled Release* 148: 135–146.

Davidson, P.M., C. Denais, M.C. Bakshi, and J. Lammerding. 2014. Nuclear deformability constitutes a rate-limiting step during cell migration in 3-D environments. *Cellular and Molecular Bioengineering* 7(3): 293–306.

Deng, Z., Z. Zhen, X. Hu, S. Wu, Z. Xu, and P.K. Chu. 2011. Hollow chitosan–silica nanospheres as pH-sensitive targeted delivery carriers in breast cancer therapy. *Biomaterials* 32(21): 4976–4986.

Dewey, C.F., Jr., S.R. Bussolari, M.A. Gimbrone, Jr., and P.F. Davies. 1981. The dynamic response of vascular endothelial cells to fluid shear stress. *Journal of Biomechanical Engineering* 103(3): 177–185.

Dhiman, H.K., A.R. Ray, and A.K. Panda. 2004. Characterization and evaluation of chitosan matrix for in vitro growth of MCF-7 breast cancer cell lines. *Biomaterials* 25(21): 5147–5154.

Ding, H.-M. and Y.-Q. Ma. 2013. Controlling cellular uptake of nanoparticles with pH-sensitive polymers. *Scientific Reports* 3: 2804.

Duan, J.-J., W. Qiu, S.-L. Xu, B. Wang, X.-Z. Ye, Y.-F. Ping, X. Zhang, X.-W. Bian, and S.-C. Yu. 2013. Strategies for isolating and enriching cancer stem cells: Well begun is half done. *Stem Cells and Development* 22(16): 2221–2239.

Egeblad, M., M.G. Rasch, and V.M. Weaver. 2010. Dynamic interplay between the collagen scaffold and tumor evolution. *Current Opinion in Cell Biology* 22: 697–706.

Elenbaas, B. and R.A. Weinberg. 2001. Heterotypic signaling between epithelial tumor cells and fibroblasts in carcinoma formation. *Experimental Cell Research* 264(1): 169–184.

Engels, B., D.A. Rowley, and H. Schreiber. 2012. Targeting stroma to treat cancers. *Seminars in Cancer Biology* 22: 41–49.

Ferrara, N. 2002. VEGF and the quest for tumour angiogenesis factors. *Nature Reviews Cancer* 2(10): 795–803.

Fischbach, C., H.J. Kong, S.X. Hsiong, M.B. Evangelista, W. Yuen, and D.J. Mooney. 2009. Cancer cell angiogenic capability is regulated by 3D culture and integrin engagement. *Proceedings of the National Academy of Sciences of the United States of America* 106(2): 399–404.

Fong, E.L.S., S.-E. Lamhamedi-Cherradi, E. Burdett, V. Ramamoorthy, A.J. Lazar, F.K. Kasper, M.C. Farach-Carson et al. 2013. Modeling ewing sarcoma tumors in vitro with 3D scaffolds. *Proceedings of the National Academy of Sciences of the United States of America* 110(16): 6500–6505.

Franco, C.A., M.L. Jones, M.O. Bernabeu, A.-C. Vion, P. Barbacena, J. Fan, T. Mathivet et al. 2016. Non-canonical Wnt signalling modulates the endothelial shear stress flow sensor in vascular remodelling. *eLife* 5(February): 1–22.

Gao, M.-Q., B.G. Kim, S. Kang, Y.P. Choi, H. Park, K.S. Kang, and N.H. Cho. 2010. Stromal fibroblasts from the interface zone of human breast carcinomas induce an epithelial-mesenchymal transition-like state in breast cancer cells in vitro. *Journal of Cell Science* 123(Pt 20): 3507–3514.

Gelain, F., D. Bottai, A. Vescovi, and S. Zhang. 2006. Designer self-assembling peptide nanofiber scaffolds for adult mouse neural stem cell 3-dimensional cultures. *PLoS ONE* 1(1): 3–44.

Gerber, P.J., C. Lehmann, P. Gehr, and S. Schürch. 2006. Wetting and spreading of a surfactant film on solid particles: Influence of sharp edges and surface irregularities. *Langmuir* 22(12): 5273–5281.

Gurski, L.A., A.K. Jha, C. Zhang, X. Jia, and M.C. Farach-Carson. 2009. Hyaluronic acid-based hydrogels as 3D matrices for in vitro evaluation of chemotherapeutic drugs using poorly adherent prostate cancer cells. *Biomaterials* 30(30): 6076–6085.

Hielscher, A.C. and S. Gerecht. 2012. Engineering approaches for investigating tumor angiogenesis: Exploiting the role of the extracellular matrix. *Cancer Research* 72: 6089–6096.

Huang, B., Z. Lei, G.-M. Zhang, D. Li, C. Song, B. Li, Y. Liu et al. 2008. SCF-mediated mast cell infiltration and activation exacerbate the inflammation and immunosuppression in tumor microenvironment. *Blood* 112(4): 1269–1279.

Huang, N.F., J. Okogbaa, J.C. Lee, A. Jha, T.S. Zaitseva, M.V. Paukshto, J.S. Sun, N. Punjya, G.G. Fuller, and J.P. Cooke. 2013. The modulation of endothelial cell morphology, function, and survival using anisotropic nanofibrillar collagen scaffolds. *Biomaterials* 34(16): 4038–4047.

Jabbari, E., S.K. Sarvestani, L. Daneshian, and S. Moeinzadeh. 2015. Optimum 3D matrix stiffness for maintenance of cancer stem cells is dependent on tissue origin of cancer cells. Edited by Adam J. Engler. *PLoS ONE* 10(7): e0132377.

Jain, R.K. 1998. The next frontier of molecular medicine: Delivery of therapeutics. *Nature Medicine* 4(6): 655–657.

Jin, S.E., H.E. Jin, and S.S. Hong. 2014. Targeted delivery system of nanobiomaterials in anticancer therapy: From cells to clinics. *BioMed Research International* 2014. 814208 (article id). doi:10.1155/2014/814208.

Kalluri, R. and M. Zeisberg. 2006. Fibroblasts in cancer. *Nature Reviews Cancer* 6(5): 392–401.

Kaplan, J.A., R. Liu, J.D. Freedman, R. Padera, J. Schwartz, Y.L. Colson, and M.W. Grinstaff. 2016. Prevention of lung cancer recurrence using cisplatin-loaded superhydrophobic nanofiber meshes. *Biomaterials* 76(January): 273–281.

Karnoub, A.E., A.B. Dash, A.P. Vo, A. Sullivan, M.W. Brooks, G.W. Bell, A.L. Richardson, K. Polyak, R. Tubo, and R.A. Weinberg. 2007. Mesenchymal stem cells within tumour stroma promote breast cancer metastasis. *Nature* 449(7162): 557–563.

Kass, L., J.T. Erler, M. Dembo, and V.M. Weaver. 2007. Mammary epithelial cell: Influence of extracellular matrix composition and organization during development and tumorigenesis. *The International Journal of Biochemistry & Cell Biology* 39(11): 1987–1994.

Kheirolomoom, A., C.-Y. Lai, S.M. Tam, L.M. Mahakian, E.S. Ingham, K.D. Watson, and K.W. Ferrara. 2013. Complete regression of local cancer using temperature-sensitive liposomes combined with ultrasound-mediated hyperthermia. *Journal of Controlled Release* 172(1): 266–273.

Kievit, F.M., S.J. Florczyk, M.C. Leung, O. Veiseh, J.O. Park, M.L. Disis, and M. Zhang. 2010. Chitosan–alginate 3D scaffolds as a mimic of the glioma tumor microenvironment. *Biomaterials* 31(22): 5903–5910.

Kleinman, H.K., M.L. McGarvey, J.R. Hassell, V.L. Star, F.B. Cannon, G.W. Laurie, and G.R. Martin. 1986. Basement membrane complexes with biological activity. *Biochemistry* 25(2): 312–318.

Kohn, J.C., D.W. Zhou, F. Bordeleau, A.L. Zhou, B.N. Mason, M.J. Mitchell, M.R. King, and C.A. Reinhart-King. 2015. Cooperative effects of matrix stiffness and fluid shear stress on endothelial cell behavior. *Biophysical Journal* 108(3): 471–478.

Kukowska-Latallo, J.F., K.A. Candido, Z. Cao, S.S. Nigavekar, I.J. Majoros, T.P. Thomas, L.P. Balogh, M.K. Khan, and J.R. Baker. 2005. Nanoparticle targeting of anticancer drug improves therapeutic response in animal model of human epithelial cancer. *Cancer Research* 65(12): 5317–5324.

Kwon, H., H.J. Kim, W.L. Rice, B. Subramanian, S.H. Park, I. Georgakoudi, and D.L. Kaplan. 2010. Development of an in vitro model to study the impact of BMP-2 on metastasis to bone. *Journal of Tissue Engineering and Regenerative Medicine* 4(8): 590–599.

Le, Q.-T., J. Harris, A.M. Magliocco, C.S. Kong, R. Diaz, B. Shin, H. Cao et al. 2009. Validation of lysyl oxidase as a prognostic marker for metastasis and survival in head and neck squamous cell carcinoma: Radiation therapy oncology group trial 90-03. *Journal of Clinical Oncology: Official Journal of the American Society of Clinical Oncology* 27(26): 4281–4286.

Lee, C.C., E.R. Gillies, M.E. Fox, S.J. Guillaudeu, J.M.J. Frechet, E.E. Dy, and F.C. Szoka. 2006. A single dose of doxorubicin-functionalized bow-tie dendrimer cures mice bearing C-26 colon carcinomas. *Proceedings of the National Academy of Sciences of the United States of America* 103(45): 16649–16654.

Lei, Y., M. Rahim, Q. Ng, and T. Segura. 2011. Hyaluronic acid and fibrin hydrogels with concentrated DNA/PEI polyplexes for local gene delivery. *Journal of Controlled Release* 153(3): 255–261.

Levental, K.R., H. Yu, L. Kass, J.N. Lakins, M. Egeblad, J.T. Erler, S.F.T. Fong et al. 2009. Matrix crosslinking forces tumor progression by enhancing integrin signaling. *Cell* 139(5): 891–906.

Li, S.D. and L. Huang. 2010. Stealth nanoparticles: High density but sheddable PEG is a key for tumor targeting. *Journal of Controlled Release* 145: 178–181.

Liao, D. and R.S. Johnson. 2007. Hypoxia: A key regulator of angiogenesis in cancer. *Cancer and Metastasis Reviews* 26: 281–290.

Lin, C.C. and A.T. Metters. 2006. Hydrogels in controlled release formulations: Network design and mathematical modeling. *Advanced Drug Delivery Reviews* 58: 1379–1408.

Liu, J., Y. Tan, H. Zhang, Y. Zhang, P. Xu, J. Chen, Y.-C. Poh, K. Tang, N. Wang, and B. Huang. 2012. Soft fibrin gels promote selection and growth of tumorigenic cells. *Nature Materials* 11(8): 734–741.

Liu, T., C. Li, H. Li, S. Zeng, J. Qin, and B. Lin. 2009. A microfluidic device for characterizing the invasion of cancer cells in 3-D matrix. *Electrophoresis* 30(24): 4285–4291.

Lu, P., V.M. Weaver, and Z. Werb. 2012. The extracellular matrix: A dynamic niche in cancer progression. *Journal of Cell Biology* 196(4): 395–406.

Maeda, H., J. Wu, T. Sawa, Y. Matsumura, and K. Hori. 2000. Tumor vascular permeability and the EPR effect in macromolecular therapeutics: A review. *Journal of Controlled Release* 65: 271–284.

Mak, I.W., N. Evaniew, and M. Ghert. 2014. Lost in translation: Animal models and clinical trials in cancer treatment. *American Journal of Translational Research* 6(2): 114–118. http://www.pubmedcentral.nih.gov/articlerender.fcgi?artid=3902 221&tool=pmcentrez&rendertype=abstract. Accessed March 23, 2016.

Menrad, H., C. Werno, T. Schmid, E. Copanaki, T. Deller, N. Dehne, and B. Brüne. 2010. Roles of hypoxia-inducible factor-1alpha (HIF-1alpha) versus HIF-2alpha in the survival of hepatocellular tumor spheroids. *Hepatology (Baltimore, MD)* 51(6): 2183–2192.

Michalet, X., F.F. Pinaud, L.A. Bentolila, J.M. Tsay, S. Doose, J.J. Li, G. Sundaresan, A.M. Wu, S.S. Gambhir, and S. Weiss. 2005. Quantum dots for live cells, in vivo imaging, and diagnostics. *Science* 307(5709): 538.

Mitragotri, S. and J. Lahann. 2009. Physical approaches to biomaterial design. *Nature Materials* 8(1): 15–23.

Miyashita, T., S. Krajewski, M. Krajewska, H.G. Wang, H.K. Lin, D.A. Liebermann, B. Hoffman, and J.C. Reed. 1994. Tumor suppressor p53 is a regulator of Bcl-2 and Bax gene expression in vitro and in vivo. *Oncogene* 9(6): 1799–1805. http://www.scopus.com/inward/record.url?eid=2-s2.0-0028335717&partnerID=tZOtx3y1. Accessed March 22, 2016.

Montet, X., R. Weissleder, and L. Josephson. 2006. Imaging pancreatic cancer with a peptide-nanoparticle conjugate targeted to normal pancreas. *Bioconjugate Chemistry* 17(4): 905–911.

Moore, A. 2004. In vivo targeting of underglycosylated MUC-1 tumor antigen using a multimodal imaging probe. *Cancer Research* 64(5): 1821–1827.

Nair, L.S. and C.T. Laurencin. 2007. Biodegradable polymers as biomaterials. *Progress in Polymer Science (Oxford)* 32: 762–798.

Nemir, S. and J.L. West. 2010. Synthetic materials in the study of cell response to substrate rigidity. *Annals of Biomedical Engineering* 38(1): 2–20.

Nguyen, D., Y.M. Akay, and M. Akay. 2016. Investigating glioblastoma angiogenesis using a 3D in vitro gelma microwell platform. *IEEE Transactions on NanoBioscience* 1241(c): 1.

Nigavekar, S.S., L.Y. Sung, M. Llanes, A. El-Jawahri, T.S. Lawrence, C.W. Becker, L. Balogh, and M.K. Khan. 2004. 3 H dendrimer nanoparticle organ/tumor distribution. *Pharmaceutical Research* 21(3): 476–483.

Ordikhani, F., Y. Kim, and S.P. Zustiak. 2015. The role of biomaterials on cancer stem cell enrichment and behavior. *JOM* 67(11): 2543–2549.

Orimo, A., P.B. Gupta, D.C. Sgroi, F. Arenzana-Seisdedos, T. Delaunay, R. Naeem, V.J. Carey, A.L. Richardson, and R.A. Weinberg. 2005. Stromal fibroblasts present in invasive human breast carcinomas promote tumor growth and angiogenesis through elevated SDF-1/CXCL12 secretion. *Cell* 121(3): 335–348.

Paliwal, S.R., R. Paliwal, H.C. Pal, A.K. Saxena, P.R. Sharma, P.N. Gupta, G.P. Agrawal, and S.P. Vyas. 2012. Estrogen-anchored ph-sensitive liposomes as nanomodule designed for site-specific delivery of doxorubicin in breast cancer therapy. *Molecular Pharmaceutics* 9(1): 176–186.

Perrault, S.D., C. Walkey, T. Jennings, H.C. Fischer, and W.C.W. Chan. 2009. Mediating tumor targeting efficiency of nanoparticles through design. *Nano Letters* 9(5): 1909–1915.

Pollard, J.W. 2004. Tumour-educated macrophages promote tumour progression and metastasis. *Nature Reviews Cancer* 4(1): 71–78.

Poon, Z., D. Chang, X. Zhao, and P.T. Hammond. 2011. Layer-by-layer nanoparticles with a pH-sheddable layer for in vivo targeting of tumor hypoxia. *ACS Nano* 5(6): 4284–4292.

Provenzano, P.P., K.W. Eliceiri, J.M. Campbell, D.R. Inman, J.G. White, and P.J. Keely. 2006. Collagen reorganization at the tumor-stromal interface facilitates local invasion. *BMC Medicine* 4(1): 38.

Rubashkin, M.G., L. Cassereau, R. Bainer, C.C. DuFort, Y. Yui, G. Ou, M.J. Paszek, M.W. Davidson, Y.-Y. Chen, and V.M. Weaver. 2014. Force engages vinculin and promotes tumor progression by enhancing PI3K activation of phosphatidylinositol (3,4,5)-triphosphate. *Cancer Research* 74(17): 4597–4611.

Saad, S., L.J. Bendall, A. James, D.J. Gottlieb, and K.F. Bradstock. 2000. Induction of matrix metalloproteinases MMP-1 and MMP-2 by co-culture of breast cancer cells and bone marrow fibroblasts. *Breast Cancer Research and Treatment* 63(2): 105–115.

Saadin, K. and I.M. White. 2013. Breast cancer stem cell enrichment and isolation by mammosphere culture and its potential diagnostic applications. *Expert Review of Molecular Diagnostics* 13(1): 49–60.

Sabeh, F., I. Ota, K. Holmbeck, H. Birkedal-Hansen, P. Soloway, M. Balbin, C. Lopez-Otin et al. 2004. Tumor cell traffic through the extracellular matrix is controlled by the membrane-anchored collagenase MT1-MMP. *Journal of Cell Biology* 167(4): 769–781.

Santoro, M., S.-E. Lamhamedi-Cherradi, B.A. Menegaz, J.A. Ludwig, and A.G. Mikos. 2015. Flow perfusion effects on three-dimensional culture and drug sensitivity of ewing sarcoma. *Proceedings of the National Academy of Sciences of the United States of America* 112: 10304–10309.

Schleich, N., C. Po, D. Jacobs, B. Ucakar, B. Gallez, F. Danhier, and V. Préat. 2014. Comparison of active, passive and magnetic targeting to tumors of multifunctional paclitaxel/SPIO-loaded nanoparticles for tumor imaging and therapy. *Journal of Controlled Release* 194: 82–91.

Schor, S.L., A.M. Schor, B. Winn, and G. Rushton. 1982. The use of three-dimensional collagen gels for the study of tumour cell invasion in vitro: Experimental parameters influencing cell migration into the gel matrix. *International Journal of Cancer* 29(1): 57–62. http://www.ncbi.nlm.nih.gov/pubmed/7061174. Accessed March 20, 2016.

Singh, P., U. Gupta, A. Asthana, and N.K. Jain. 2008. Folate and folate-PEG-PAMAM dendrimers: Synthesis, characterization, and targeted anticancer drug delivery potential in tumor bearing mice. *Bioconjugate Chemistry* 19(11): 2239–2252.

Sodek, K.L., T.J. Brown, and M.J. Ringuette. 2008. Collagen I but not matrigel matrices provide an MMP-dependent barrier to ovarian cancer cell penetration. *BMC Cancer* 8: 223.

Soman, P., J.A. Kelber, J.W. Lee, T.N. Wright, K.S. Vecchio, R.L. Klemke, and S. Chen. 2012. Cancer cell migration within 3D layer-by-layer microfabricated photocrosslinked PEG scaffolds with tunable stiffness. *Biomaterials* 33(29): 7064–7070.

Subbiah, V., P. Anderson, A.J. Lazar, E. Burdett, K. Raymond, and J.A. Ludwig. 2009. Ewing's sarcoma: Standard and experimental treatment options. *Current Treatment Options in Oncology* 10(1–2): 126–140.

Tamura, M., F. Yanagawa, S. Sugiura, T. Takagi, K. Sumaru, and T. Kanamori. 2015. Click-crosslinkable and photodegradable gelatin hydrogels for cytocompatible optical cell manipulation in natural environment. *Scientific Reports* 5(October): 15060 (Nature Publishing Group).

Thevenot, P., W. Hu, and L. Tang. 2008. Surface chemistry influences implant biocompatibility. *Current Topics in Medicinal Chemistry* 8(4): 270–280.

Torchilin, V.P. 2010. Passive and active drug targeting: Drug delivery to tumors as an example. *Handbook of Experimental Pharmacology* 197: 3–53.

Vered, M., D. Dayan, R. Yahalom, A. Dobriyan, I. Barshack, I.O. Bello, S. Kantola, and T. Salo. 2010. Cancer-associated fibroblasts and epithelial-mesenchymal transition in metastatic oral tongue squamous cell carcinoma. *International Journal of Cancer [Journal International Du Cancer]* 127(6): 1356–1362.

Voinea, M. and M. Simionescu. 2002. Designing of "Intelligent" liposomes for efficient delivery of drugs. *Journal of Cellular and Molecular Medicine* 6(4): 465–474.

Wang, H.B., M. Dembo, and Y.L. Wang. 2000. Substrate flexibility regulates growth and apoptosis of normal but not transformed cells. *American Journal of Physiology—Cell Physiology* 279(5): C1345–C1350.

Weinberg, R.A. 1994. Oncogenes and tumor suppressor genes. *CA: A Cancer Journal for Clinicians* 44(3): 160–170.

Wolf, K., M. te Lindert, M. Krause, S. Alexander, J. Te Riet, A.L. Willis, R.M. Hoffman, C.G. Figdor, S.J. Weiss, and P. Friedl. 2013. Physical limits of cell migration: Control by ECM space and nuclear deformation and tuning by proteolysis and traction force. *The Journal of Cell Biology* 201(7): 1069–1084.

Wu, Y., B. Guo, and G. Ghosh. 2015. Differential effects of tumor secreted factors on mechanosensitivity, capillary branching, and drug responsiveness in PEG hydrogels. *Annals of Biomedical Engineering* 43(9): 2279–2290.

Xu, F. and K.J.L. Burg. 2007. Three-dimensional polymeric systems for cancer cell studies. *Cytotechnology* 54(3): 135–143.

Xu, X., A.K. Jha, D.A. Harrington, M.C. Farach-Carson, and X. Jia. 2012. Hyaluronic acid-based hydrogels: From a natural polysaccharide to complex networks. *Soft Matter* 8(12): 3280.

Yang, K., J. Wan, S. Zhang, B. Tian, Y. Zhang, and Z. Liu. 2012. The influence of surface chemistry and size of nanoscale graphene oxide on photothermal therapy of cancer using ultra-low laser power. *Biomaterials* 33(7): 2206–2214.

Yang, X., S.K. Sarvestani, S. Moeinzadeh, X. He, and E. Jabbari. 2013. Effect of CD44 binding peptide conjugated to an engineered inert matrix on maintenance of breast cancer stem cells and tumorsphere formation. Edited by Ming Tan. *PLoS ONE* 8(3): e59147.

Yang, Y.L., S. Motte, and L.J. Kaufman. 2010. Pore size variable type I collagen gels and their interaction with glioma cells. *Biomaterials* 31(21): 5678–5688.

Yeung, T., P.C. Georges, L.A. Flanagan, B. Marg, M. Ortiz, M. Funaki, N. Zahir, W. Ming, V. Weaver, and P.A. Janmey. 2005. Effects of substrate stiffness on cell morphology, cytoskeletal structure, and adhesion. *Cell Motility and the Cytoskeleton* 60(1): 24–34.

Yhee, J.Y., S. Lee, and K. Kim. 2014. Advances in targeting strategies for nanoparticles in cancer imaging and therapy. *Nanoscale* 6(22): 13383–13390.

Yokoyama, M. 2005. Drug targeting with nano-sized carrier systems. *Journal of Artificial Organs* 8(2): 77–84.

Yu, M.K., J. Park, and S. Jon. 2012. Targeting strategies for multifunctional nanoparticles in cancer imaging and therapy. *Theranostics* 2: 3–44.

Zawaneh, P.N., S.P. Singh, R.F. Padera, P.W. Henderson, J.A. Spector, and D. Putnam. 2010. Design of an injectable synthetic and biodegradable surgical biomaterial. *Proceedings of the National Academy of Sciences of the United States of America* 107(24): 11014–11019.

Zhou, Z. and Z.-R. Lu. 2013. Gadolinium-based contrast agents for magnetic resonance cancer imaging. *Wiley Interdisciplinary Reviews: Nanomedicine and Nanobiotechnology* 5(1): 1–18.

5 The Cell as an Inspiration in Biomaterial Design

Helim Aranda-Espinoza and Katrina Adlerz

CONTENTS

5.1 INTRODUCTION

The diversity of life arises from genetic information that has evolved through millions of years to meet the survival needs of different species. This genetic material is stored in single cells; therefore, the cell is considered the minimal unit in a living system [1]. Living organisms can be as simple as a single cell or as complex as the human body, which is made up of 10^{13} cells [2]. Despite the vast number of cells in humans, there are only around 200 different types of cells. These cells have a large diversity of structure and function from the simplest cell, the red blood cell, to the highly complex neuron [3]. Within this diversity, however, cells have similar structural components like the plasma membrane that encloses the cell and the cytoskeleton that allows cells to organize their internal components and carry out sophisticated biological functions such as mechanically adapting to their environment, migrating, growing, and dividing [3]. Another similarity among cells is that all cells store their hereditary information in the heteropolymer chain called DNA, which, after being transcribed to RNA, is then translated to the proteins that carry out cellular functions [4].

    The cell and its different components must follow the laws of physics and chemistry but its unique structure and composition allow for a myriad of functions [5].

Biologists, physicists, and chemists have all studied the properties of cellular components to understand how the cell organizes and executes its many jobs. These studies also revealed how cells have evolved to solve many engineering-related problems like self-assembly, self-healing, and tolerating or adapting to external stimuli. These properties are highly desirable in building biomaterials, which can be defined as materials that interact with biological systems in the course of a treatment or a diagnostic. Therefore, the components of the cell have been used both as an inspiration and as a raw material in the development and synthesis of a range of biomaterials with specific mechanical properties.

This chapter will offer a summary of cellular components, including the cytoskeleton, the plasma membrane, and DNA, and explore how each of these has been utilized in the design and production of biomaterials. Furthermore, we will review work that has used whole cells to design biomaterials with specific biological functions, making use of the mechanical properties of cells.

5.2 MEMBRANES

5.2.1 Phospholipid Bilayer Structure

Membranes are 5–10 nm layers that enclose cells and certain organelles to create an environment for the many chemical reactions that occur inside the cell and organelles. A membrane is about 50% lipids and 50% proteins [6]. The majority of the lipids are amphiphilic phospholipids that have a polar head and a hydrophobic hydrocarbon tail. These amphiphilic lipids self-assemble into bilayers in aqueous environments with the heads exposed and tails inside. While water is able to move through lipid bilayers, other ions and most polar molecules are not able to diffuse across the membrane. Certain transmembrane proteins allow specific molecules and atoms to be transported through, allowing homeostasis between the inside and outside of a cell. The ability of phospholipids to self-assemble into closed structures like spherical vesicles and their selective permeability has been used in the design of biomaterials. Membrane phospholipids have been used to create enclosed vesicles that can carry drugs, DNA, siRNA, or molecules for detection. The self-assembly principles of the cell membrane have also been mimicked to create synthetic amphiphiles that can also form vesicles as explained here.

5.2.2 Liposomes for Drug Delivery

The unique nature of the lipid bilayer has inspired a class of drug delivery vehicles called liposomes. Liposomes are spherical vesicles that self-assemble from isolated phospholipids in aqueous environments (see Figure 5.1). Liposomes can be unilamellar, having only one lipid bilayer, or multilamellar, with several concentric lipid bilayers. They can also be tailored to a range of sizes. Drugs can be encapsulated inside the aqueous environment of the liposome or in the hydrophobic portion of the membrane. One advantage of liposomes is that they can be modified. For example, liposomes can incorporate targeting moieties like antibodies that bind to specific ligands on target cells. Liposomes have also been modified by adding polyethylene glycol (PEG), which increases the circulation time of injected liposomes in the blood and slows clearing by liver and spleen cells, therefore allowing more drug to be delivered [7]. The first Food and Drug Administration–approved treatment using liposomes was to treat cancer with the drug doxorubicin encapsulated in a liposome composed of phosphatidylcholine, cholesterol, and PEG. The liposome-formulated drug, called Doxil, remains in circulation longer than free doxorubicin, which allows for greater accumulation and release of the drug at the tumor [8,9]. The amount of work with liposomes is very extensive and difficult to fully cover here; the interested reader is pointed to the following reviews [10–12].

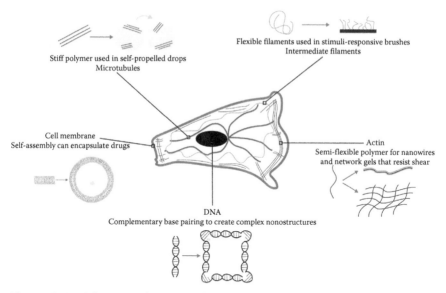

Flexible filaments used in stimuli-responsive brushes
Intermediate filaments

Stiff polymer used in self-propelled drops
Microtubules

Cell membrane
Self-assembly can encapsulate drugs

Actin
Semi-flexible polymer for nanowires
and network gels that resist shear

DNA
Complementary base pairing to create complex nonostructures

Figure 5.1 Schematic of a mammalian cell with the cellular components that have been used in biomaterial design highlighted as follows: actin in red, intermediate filaments in light blue, microtubules in green, and nucleus in navy.

A desire to improve on the mechanical properties and stability of liposomes drove the search for new materials that would be more stable and mechanically sound, while still being able to self-assemble into vesicles (see Figure 5.1). Polymer vesicles or polymersomes were developed to meet these needs [13].

5.2.3 Polymer Vesicles

While liposomes utilize phospholipids as building materials, recent research has also investigated the use of polymers in the preparation of vesicles for drug delivery and diagnosis. These structures, named polymersomes, still contain hydrophobic and hydrophilic components, but they are made of block copolymers that have a much higher molecular weight than lipids. This increased molecular weight is partly responsible for polymersomes' larger strength and stability [13]. In fact, the electromechanical properties of polymersomes have been shown to be up to a magnitude higher than liposomes [14]. Polymersomes are also more amenable to modifications to become stimuli-responsive. For example, temperature-sensitive and light-sensitive polymersomes are being developed [15]. Another interesting modification is the use of pH-sensitive diblock copolymers; in this case, the polymersome forms in a neutral pH and dissolves completely in the acidic endosome [16]. Polymer vesicles have also been designed for intracellular drug delivery using disulfide block copolymers [17]. The vesicle protects its contents outside the cell, but as it is uptaken by endocytosis, the vesicle is ruptured in the early endosome releasing its contents.

Because polymersomes are made by carefully controlling the molecular weight ratio of hydrophobic-to-hydrophilic moieties, which basically controls the packing factor [18], it is possible to manipulate the packing factor to make micelles rather than vesicles [19]. These micelles have been shown to circulate 10 times longer than vesicles made with the same chemistry and have similar drug delivery capabilities [20].

Similar to liposomes, the use of polymersomes has exploded and many other types of copolymers have been designed with a variety of uses; the curious reader is invited to read the following interesting reviews [12,21–24].

Investigators have also created amphiphilic molecules with peptides [25–29] that can be used for targeted drug delivery and diagnostics. In principle, these vesicles are safer than polymersomes and liposomes since they do not require artificial polymers like PEG to stabilize the vesicle [30]. They are also more stable at high temperatures than liposomes.

Liposomes, polymersomes, and peptosomes, all inspired by the cell membrane, hold great promise in more efficient, targeted drug delivery, that means less drug would have to be delivered systemically, therefore limiting side effects. Besides therapeutics, these vesicles are also being used for diagnostics. In fact, they can be used for both applications simultaneously, creating carriers for theranostic purposes.

5.3 CYTOSKELETON

The cytoskeleton is largely responsible for the mechanical properties and spatial organization of animal cells (see Figure 5.1). Its roles include changing the shape of the cell, coordinating migration, imparting mechanical resistance to deformation, and cell division. The cytoskeleton is a collection of three main proteins classified by the diameter of their filaments. Semi-flexible actin filaments, the smallest at around 7 nm, are mainly involved in cell migration. Flexible intermediate filaments provide mechanical strength to the cell. Microtubules with a 25 nm diameter are the largest and stiffest filament of the cytoskeleton and are involved in directing intracellular transport and are major components of cilia and flagella [2]. The unique mechanical properties of the different cytoskeletal filaments make each suited for a different type of biomaterial with specific properties and applications. Weak, noncovalent bonds and hydrogen interactions hold together each of the three components of the cytoskeleton, allowing for rapid assembly and disassembly, which in turn allows the cytoskeleton to adapt to external stimuli, another desirable property in biomaterial design.

While liposomes have been in development since the 1960s [31], it is more recently, only since the 1990s, that cytoskeletal elements have been investigated as biomaterials [32]. Nevertheless, recent research has led to a better understanding of the cytoskeleton, which has subsequently been utilized in the development of biomaterials. An overview of the structure and properties of each of the three main cytoskeletal proteins is detailed here, as well as examples of biomaterials inspired by the different components of the cytoskeleton.

5.3.1 Actin Cytoskeleton

5.3.1.1 Actin Structure

Microfilaments, also called actin filaments, are one of the three main components of the cytoskeleton and are made up of the protein actin. Actin is found in the cell in its monomer form globular actin (G-actin) or organized into linear filaments (F-actin). F-actin is a flexible polymer found at the cell cortex under the plasma membrane to provide mechanical support to the cell as well as throughout the cell where it is involved in cell migration, cell contractility, and cytokinesis [2].

Actin filaments can organize into linear bundles, two-dimensional networks, or 3D gels. Specific nucleating proteins are responsible for initiating these different structures. For example, formins associate with the fast-growing end of actin and allow new monomers to attach to the end, resulting in straight, unbranched filaments. These filaments are then cross-linked by various bundling proteins into linear bundles, which can result in noncontractile bundles, contractile stress fibers, or the tightly packed bundles seen in epithelial cells' microvilli.

Alternatively, actin-related protein (ARP) complexes associate with the side of an already-assembled actin filament and nucleate actin monomers into a new filament there, resulting in a branched network of actin. Filamin is an actin-associated protein that holds together two actin filaments at right angles forming a loose, highly viscous active gel associated with lamellipodia used in cell migration. Alternatively, spectrin forms a hexagonal mesh binding actin filaments together into a stiff 3D web and is concentrated underneath the plasma membrane forming the actin cell cortex. Actin polymerization to form filaments and the filaments' ability to form gels has captured the imagination of scientists to create interesting biomaterials like the reinforced liposomes and actin gels described here.

5.3.1.2 Actin in Liposomes

The role of the actin cortex in stabilizing the cell membrane has been employed in the development of liposomal drug delivery vehicles. Filamentous actin was incorporated into liposomes by first encapsulating G-actin into the center of liposomes. The liposomes were then put into a high ionic strength buffer, which promoted actin polymerization into F-actin. The resulting actin filaments formed a thin cortex beneath the membrane [33]. These actin-containing liposomes were more mechanically stable and could be utilized as drug delivery vehicles that are flexible enough to transverse capillaries yet strong enough to survive shear stresses present in the bloodstream.

In another application of actin in liposomes, the effects of the actin-binding protein filamin and an actin-severing protein, gelsolin, were investigated on liposome formation. G-actin was encapsulated into liposomes with either filamin or gelsolin, then the potassium chloride concentration was increased to promote the polymerization of actin filaments. Liposomes with polymerized actin filaments were slightly deformed, as actin filaments had to bend inside the liposomes. With a high ratio of gelsolin to actin, however, the length of the actin filaments decreased and the liposomes returned to a spherical shape. The addition of filamin resulted in a smoother surface [34]. These experiments illustrate how knowledge of the specific proteins involved in actin structure has been utilized in biomaterial design and could influence the development of a drug delivery system.

5.3.1.3 Actin Gels

Actin can be polymerized not just inside the cell and membranes, but under the correct conditions, in vitro as well (see Figure 5.1, in red). The mechanical properties of actin make it an interesting material to synthesize and study [35]. Actin is a semi-flexible polymer and can be described by its persistence length, which describes the length before random thermal fluctuations cause it to bend. Actin has a persistence length of around 20 μm. Another unique property of the cytoskeleton and actin are that they strain harden [36]. When a mechanical stress is applied to a cell, the cytoskeleton's elasticity increases. These desirable properties have motivated the study of actin as a biomaterial. The strain hardening behavior was seen in vitro with gels of F-actin [36]. It was also found that α-actinin, a member of the spectrin family of actin cross-linking proteins enhanced the strain hardening of the F-actin, allowing the gels to withstand even larger deformations [37].

5.3.1.4 Active Actin Gels

Another attractive property of the actin cytoskeleton for biomaterials is its ability to provide mechanical integrity while being able to completely remodel. This property is, in part, due to molecular motors that move along cytoskeletal elements. The molecular motor myosin binds to actin filaments and moves toward the plus ends using ATP hydrolysis. The myosin II motor is involved in generating forces for muscle contraction, and there are other myosins that function to carry cargo across the cell.

Gels made up of actin and myosin are called active gel systems because even in vitro the myosin walks along the actin. These systems have been useful as a tool to study active force generation by motors and force dissipation in the viscoelastic cytoskeleton. A review on the physics of active gels can be found in Reference 38. For example, Kohler, Schaller, and Bausch studied a gel of myosin II, actin filaments, and the cross-linking molecule fascin, often found in cell membrane ruffles. They observed the growth properties and dynamics of clusters of actin networks and found that they depend on the density of actin filaments and molecular motors [39]. The transport dynamics they saw were similar to what has been seen in vivo. A better understanding of the structure of actin and motors could be utilized in various nanotechnologies.

For example, actin and myosin are being explored for their potential usefulness as nanotransporters for applications in nanoelectronics and biosensors; interested readers are invited to see References 40 and 41 for a more detailed review on motor proteins in nanotechnology. Patolsky, Weizmann, and Willner made actin-based metallic nanowires by reacting polymerized F-actin with gold nanoparticles resulting in 2–3 µm long conductive wires (see Figure 5.1, in gold and red). These wires were then plated on a surface of myosin motors, and after the addition of ATP about 30% of the nanowires moved on the surface [42].

Active gel systems could be used to fabricate precise nanoscale materials or to transport specific proteins in biosensor applications. A better understanding of how to move motors along specific routes will be key to exploiting these biological motors in nanotechnology.

5.3.2 Intermediate Filaments

5.3.2.1 Intermediate Filament Structure

All vertebrates have intermediate filaments, although they are most prominent in cells that have to withstand high mechanical stresses. They play an important role in imparting mechanical strength to cells and tissues. Compared to actin and microtubules, the other major components of the cytoskeleton, intermediate filaments have more diversity. Instead of being made up of one type of protein, there are many different monomers that can make up intermediate filaments, and the composition depends on the cell type. For example, keratin monomers form the intermediate filaments found in human epithelial cells, and the cross-linked keratin networks give strength to hair and nails. Vimentin filaments are a second type of intermediate filament that are found in mesenchymal cells and help anchor organelles in the cytosol and maintain cell integrity. A third example of intermediate filaments is neurofilaments that are found along axons and provide structural support to the axon [2]. Neurofilaments are made up of three subunit proteins classified by their molecular weight: low, medium, and high. The subunits have the same basic structure but the lengths of their sidearms differ. The low molecular weight subunit has the shortest sidearm, and the high molecular weight subunit has the longest. These sidearms are thought to mediate the spacing between filaments. Overaccumulation of neurofilaments can block the transport of proteins down the axon and is seen in diseases like dementia, Parkinson's, and amyotrophic lateral sclerosis [43].

5.3.2.2 Intermediate Filament Gels Respond to External Stimuli

The different subunits and unique network of intermediate filaments have motivated studies of neurofilaments as possible stimuli-responsive biomaterials. Neurofilaments can be reconstituted as in vitro gels where the responses of the gels to external stimuli can be measured. Beck et al. found that when an osmotic pressure was applied, the neurofilament gels transitioned from an expanded to a condensed state, dependent on the subunit expression [43].

Other researchers capitalized on the behavior of these cytoplasmic gels and explored neurofilament proteins as a biosynthetic coating. Polymer brushes minimize nonspecific interactions and can respond to stimuli like osmotic pressure and pH that could be useful in sensors and drug-release systems (see Figure 5.1, in light blue). While many polymer brushes are synthetic peptides, this research showed that neurofilament-inspired brushes are a biomimetic alternative. High molecular weight neurofilament proteins were attached to a substrate and tested at various pH conditions showing that brush height depended on both pH and ionic strength [44].

5.3.3 Microtubules

5.3.3.1 Microtubule Structure

Microtubules are the largest of the three major cytoskeletal filaments, and they are also the stiffest with a persistence length of several millimeters [2]. They are made up of alpha- and beta-tubulin, which form tubulin heterodimers that assemble into protofilaments; 13 protofilaments arrange into hollow cylinders to make up the microtubule filament. A variety of microtubule-associated proteins (MAPs) then bind to some microtubules to stabilize the structures discouraging disassembly, and are responsible for creating a network of microtubules. MAP2, for example, helps to form bundles of widely spaced, stable microtubules. Contrary to actin filaments, microtubules can polymerize and depolymerize in a rather fast way, this phenomenon is known as dynamic instability, and it finds uses in cell division. Because of the importance of microtubules in cell division, they are perfect targets to control unregulated cell division, that is, cancer. Microtubules in the cell form the highways through which many proteins and vesicles are delivered to the periphery. In neurons, for example, the axons are mostly microtubules and intermediate filaments: the intermediate filaments are used for axonal caliber, while the microtubules are used to transport vesicles to the tip of the axon and back. Mechanically, microtubules are the stiffest filament in the cell and thus they are in charge of supporting the tension generated by actomyosin. Furthermore, the stiff microtubules can be organized in bundles to create cilia or flagella that can oscillate with the help of molecular motors, as will be discussed here.

5.3.3.2 Microtubule Gel

Microtubule gels can be made in bulk in vitro by taking advantage of the self-assembly of tubulin dimers [35]. Sano et al. utilized the ability of microtubules to quickly polymerize and depolymerize to create a thermoresponsive gel. First, tubulin was heated to allow for polymerization into microtubule filaments. The filaments were then mixed with bis-NHS PEG, which cross-linked the filaments into a gel. At physiological temperatures the microtubules remained polymerized and the gel solidified. When the gel was put on ice, the tubulin depolymerized and became a sol [45]. This process was found to be repeatable with the same gel. This type of biocompatible, thermoresponsive gel is useful in biomedical applications because it is easier to deliver to the body as a solution than a gel. It may be useful as a drug delivery system or in tissue engineering applications.

5.3.3.3 Active Microtubule Gels

Similar to how myosin can move along actin filaments, motor proteins also walk along microtubules. Instead of just one motor protein though, microtubules have two different classes of motors: kinesin that moves toward the growing, plus end of microtubules, and dynein that moves toward the minus end, which is typically anchored at the microtubule organizing center. These motor proteins can bind other microtubules, organelles, or vesicles to transport along a microtubule.

In order to better understand microtubule motor dynamics, in vitro systems of microtubules and kinesins were mixed and their collective behavior was analyzed. Surrey, Nédélec, Leibler, and Karsenti found that at low motor concentrations the microtubules were randomly organized, but as the concentration increased, the microtubules formed vortices at intermediate motor concentrations and formed asters at high motor protein concentrations [46]. These experiments showed that in vivo dynamics could be observed with in vitro systems.

Sanchez, Welch, Nicastro, and Dogic extended this idea to recreate the metachronal waves seen in beating cilia. Cilia and flagella are organelles made up of microtubules and dynein. Flagella are found on sperm and are responsible for their movement. Cilia can be immotile like those found on single cells or motile like the many found on cells in the respiratory tract that help to move mucus out of the lungs. While actual cilia contain more than 600 different proteins, a similar synchronized beating of microtubules was observed with a simplified system. A complex of biotin-labeled kinesin motors bound to multimeric streptavidin could bind and walk on multiple taxol-stabilized microtubules. PEG coils were then added, which induces microtubule attraction through depletion forces: that is, the microtubules came together because the gap between filaments is not big enough for the PEG, creating an osmotic differ-ence that attracts the filaments to each other [47]. Surprisingly, when these microtu-bule bundle structures attached to a wall they spontaneously beat similar to cilia.

This same system, which recapitulates the biological phenomena of cilia beating, was also investigated for its use as an active material. When the microtubule, kinesin, PEG system was confined in small lipid vesicles around 30 μm in diameter, the gel adsorbed to the inner surface of the droplet and the droplet was able to move along a surface, it was measured to move around 250 μm in 35 min [48] (see Figure 5.1, in green). Further investigations could look into whether this movement can be directed and controlled, if so these droplets could be used as transporters that do not require an outside energy source in various nanotechnology applications like biosensors [41].

5.4 DNA

5.4.1 DNA Structure and Complementary Base Pairing

DNA is the carrier of genetic information for all forms of life. DNA is a nucleic acid made up of two long polynucleotide chains. There are four nucleotides A, C, G, and T that form the polynucleotide backbone through covalent linkages longitudinally. The nucleotide bases then form hydrogen bonds with another polynucleotide backbone to create a double-stranded chain, which twists into a double helix. The A nucleotide always bonds with T, while C and G always bond with each other [2]. This is called complementary base pairing and has been exploited to create new, precisely designed materials.

In vitro, double helixes form from mixtures of single-stranded DNA in a process called hybridization. In this process, double helixes are first denatured by heat into single strands. When these single strands are mixed in a test tube, random collisions take place and occasionally these collisions result in a DNA strand binding to a short sequence of another strand with complementary base pairing called helix nucle-ation. From this point, a rapid zippering proceeds along the rest of the two strands creating a complete double helix. A single strand is able to find its complementary sequence even among millions of nonmatching sequences, illustrating the robustness of complementary base pairing. This specificity inspired scientists who used DNA's properties of self-assembly and complementary base pairing to create well-defined structures like the branched DNA gels and DNA origami described here (see Figure 5.1). In addition to complementary base pairing, the molecular machinery involved in the DNA process, like DNA replication, has also been exploited in biomaterial design.

5.4.2 DNA Replication

DNA replication is necessary during cell division so that each daughter cell gets a complete copy of DNA from the mother cell. Each of the double strands of the DNA helix serves as a template for the enzyme DNA polymerase to create complementary strands. DNA polymerase only works in one direction along DNA from the 5' end, the end with a phosphate, to the 3' end, the end with a hydroxyl. For the leading strand, the DNA polymerase moves along the unwinding DNA, polymerizing a continuous strand of DNA. As the DNA polymerase moves along the lagging strand, however, it polymerizes short sequences in the 5'–3' direction and an additional enzyme called DNA ligase pieces together these Okazaki fragments.

5.4.3 Branched DNA Gels Utilize DNA Ligase

Um et al. created macroscopic gels made out of specific DNA molecules that are pieced together with the DNA ligase enzyme used in DNA replication. They created various branched DNA structures shaped as an X, Y, and T and used them as monomer building blocks. The shapes had sticky ends that were complementary to each other and could be cross-linked by adding DNA ligase to create a gel network. The different-shaped monomers gave the gels distinct properties like tensile modulus and degradation [49]. The material is being investigated as a biodegradable, biocompatible, inexpensive material for drug delivery. As a proof of principle, insulin was encapsulated into the DNA gel and it released slowly over approximately 12 days, depending on the shape of the branched DNA monomer [49].

5.4.4 DNA Double Crossover Molecules as Building Blocks

Seeman and colleagues were among the first to build nanostructures with DNA [50]. Their methods borrowed double crossover molecules, another aspect of biological DNA, to create rigid, crystalline structures. While branched DNA molecules have an ill-defined structure because the branch point angles can change, double crossover molecules are more defined. Double crossover structures are seen in the cell during DNA repair and during meiosis when additional genetic variability is introduced. Briefly, in either case, both strands of a DNA double helix are broken. This helix is in close contact with an intact sister chromatid. Each side of the severed DNA strand finds its complement in the intact strand and two branch points are formed where the intact strand and the severed strand cross. DNA is then synthesized and ligated onto the severed strand based on the template of the intact strand. Winfree et al. designed and synthesized various double crossover molecules. The molecules were designed according to the principles of mathematic tiling so that the molecules could fit together in a 2D crystalline structure. Each molecule was around $2 \times 4 \times 15$ nm but could self-assemble to form crystals as big as 2×8 μm [50].

5.4.5 DNA Origami

DNA origami expands upon previous work using DNA as a building block and allows for more complex designs. Rothemund pioneered the technique that uses a computer program to design short DNA strands, named "staples," that are complementary to specific points on a long single strand of DNA. The staples are mixed with the long DNA, heated, and allowed to cool. As the mixture cools, one side of a staple binds to a section of the long DNA while the staple's other side binds somewhere else along the single strand, bringing these two sections close to each other. After all the staple strands have bound to the long DNA strand, the DNA will resemble the user-inputted shape [51]. DNA triangles, smiley faces [51], and even complex curves [52] and boxes that can open and shut [53] have been built based on this technique. DNA origami uses relatively cheap materials and takes advantage of biology's complementary base pairing to build extremely small, precise nanostructures.

Investigations into using DNA origami as a drug delivery system are already underway. For example, the drug Dox was loaded into DNA origami nanostructures through intercalation [54,55]. Cells took up the nanostructures, leading to the death of cancerous cells that were not killed by free Dox, suggesting that the system could be used to target drug-resistant cells [55].

Further work optimizing the use of DNA in nanomaterials is ongoing as is the development of specific applications in nanotechnology. There are a number of review articles that more thoroughly describe the progression and current state of the technology [56,57].

5.5 CELL-BASED BIOMATERIALS

We have described how cellular components have been utilized to create new biomaterials with specific mechanical properties. It is also possible to use the physiological functions of cells to create biomaterials that can carry out mechanical work. In particular, cardiomyocytes were used by Tanaka et al. to create micropumps. The force generated by cardiomyocytes cultured on elastic sheets allowed the researchers to move fluid at 0.24 μL/min [58]. Similarly, Choi et al. combined cardiomyocytes with piezoelectric materials to generate electric currents. The authors were able to generate an output voltage measured between 1.48 and 4.19 mV [59]. Using cells as biomaterials was taken to new dimensions when researchers created a biological robot that swims like a stingray. Park et al. built a device made up of a gold skeleton encapsulated in an elastomer body that was seeded with cardiomyocytes. The cardiomyocytes were genetically engineered to respond to light. The engineered robot was then able to respond to light cues and swim in a similar manner to a real ray [60]. This research could lead to the creation of cell-based organoids, which could help in the design of artificial organs.

5.6 CONCLUSIONS AND FUTURE RESEARCH WITH BIOINSPIRED MATERIALS

Biology has served as an inspiration in the design of many materials. The inspiration for Velcro, for example, was inspired by plant burrs that clung to the inventor's pants during a hike. Similarly, reflectors in the road are called cat's eye because they were modeled after the way that cats' eyes reflect light.

Biomaterials that interact with the body as drug delivery systems, biosensors, or diagnostic tools have specific design parameters and unique challenges like biocompatibility and degradation. As biology and specifically the cell and its components have been studied and better understood, their unique properties have served as inspiration for addressing these challenges in the design of biomaterials. The cell, therefore, has influenced biomaterial design in properties like self-assembly, stimuli-responsiveness, and mechanical stability. Likewise, the creation and study of biomaterials has offered insights into biology, which in turn again influences biomaterial design and development.

We have discussed how specific components of the cell have served as inspiration or been used as raw material in a variety of biomaterials. For example, the ability of membranes to self-assemble into closed vesicles has been utilized in drug delivery. The specific properties of each of the cytoskeletal elements have given rise to a variety of biomaterials like flexible brushes of intermediate filaments, actin nanowires, and stiff microtubules in self-propelled drops. DNA's complementary base pairing, meanwhile, has been exploited in the construction of precise nanostructures.

While liposomes are already in the clinic as a drug delivery vehicle, many of the other materials are still being investigated in the laboratory for clinical applications like drug delivery systems, diagnostic devices, and nanotechnology. A desire to better understand how to utilize the unique properties of cellular components has given rise to an exciting field of study in bioinspired materials research.

REFERENCES

1. Lodish H, Berk A, Zipursky SL, Matsudaira P, Baltimore D, Darnell J. *Molecular Cell Biology*. 4th edn. New York: W. H. Freeman; 2000.

2. Alberts B, Johnson A, Lewis J, Raff M, Roberts K, Walter P. *Molecular Biology of the Cell*. 5th edn. New York: Garland Science; 2007. 1392pp.

3. Bray D. *Cell Movements: From Molecules to Motility*. 2nd edn. New York: Garland Science; 2000. 392pp.

4. Calladine CR, Drew H, Luisi B, Travers A. *Understanding DNA: The Molecule and How It Works*. 3rd edn. San Diego, CA: Academic Press; 2004. 352pp.

5. Phillips R, Kondev J, Theriot J, Garcia H. *Physical Biology of the Cell*. 2nd edn. London, U.K.: Garland Science; 2012. 1057pp.

6. Gennis RB. *Biomembranes: Molecular Structure and Function*. New York: Springer Science & Business Media; 1989. 549pp.

7. Klibanov AL, Maruyama K, Torchilin VP, Huang L. Amphipathic polyethyleneglycols effectively prolong the circulation time of liposomes. *FEBS Lett*. July 30, 1990;268(1):235–237.

8. Gabizon A, Catane R, Uziely B, Kaufman B, Safra T, Cohen R et al. Prolonged circulation time and enhanced accumulation in malignant exudates of doxorubicin encapsulated in polyethylene-glycol coated liposomes. *Cancer Res*. Feb 15, 1994;54(4):987–992.

9. Muggia FM, Hainsworth JD, Jeffers S, Miller P, Groshen S, Tan M et al. Phase II study of liposomal doxorubicin in refractory ovarian cancer: Antitumor activity and toxicity modification by liposomal encapsulation. *J Clin Oncol*. Mar 1, 1997;15(3):987–993.

10. Peer D, Karp JM, Hong S, Farokhzad OC, Margalit R, Langer R. Nanocarriers as an emerging platform for cancer therapy. *Nat Nanotechnol*. 2007;2(12):751–760.

11. Torchilin VP. Recent advances with liposomes as pharmaceutical carriers. *Nat Rev Drug Discov*. 2005;4(2):145–160.

12. Müller LK, Landfester K. Natural liposomes and synthetic polymeric structures for biomedical applications. *Biochem Biophys Res Commun*. 2015;468(3):411–418.

13. Discher BM, Won Y-Y, Ege DS, Lee JC-M, Bates FS, Discher DE et al. Polymersomes: Tough vesicles made from diblock copolymers. *Science*. May 14, 1999;284(5417):1143–1146.

14. Aranda-Espinoza H, Bermudez H, Bates FS, Discher DE. Electromechanical limits of polymersomes. *Phys Rev Lett*. Oct 24, 2001;87(20):208301.

15. Meng F, Zhong Z, Feijen J. Stimuli-responsive polymersomes for programmed drug delivery. *Biomacromolecules*. 2009;10(2):197–209.

16. Lomas H, Canton I, MacNeil S, Du J, Armes SP, Ryan AJ et al. Biomimetic pH sensitive polymersomes for efficient DNA encapsulation and delivery. *Adv Mater*. 2007;19(23):4238–4243.

17. Cerritelli S, Velluto D, Hubbell JA. PEG-SS-PPS: Reduction-sensitive disulfide block copolymer vesicles for intracellular drug delivery. *Biomacromolecules*. 2007;8(6):1966–1972.

18. Israelachvili JN. *Intermolecular and Surface Forces*, Revised 3rd edn. Burlington, MA: Academic Press; 2011. 704pp.

19. Dalhaimer P, Bates FS, Discher DE. Single molecule visualization of stable, stiffness-tunable, flow-conforming worm micelles. *Macromolecules*. 2003;36(18):6873–6877.

20. Geng YAN, Dalhaimer P, Cai S, Tsai R, Tewari M, Minko T et al. Shape effects of filaments versus spherical particles in flow and drug delivery. *Nat Nanotechnol*. 2007;2(4):249–255.

21. Du J, O'Reilly RK. Advances and challenges in smart and functional polymer vesicles. *Soft Matter*. 2009;5(19):3544–3561.

22. Blanazs A, Armes SP, Ryan AJ. Self-assembled block copolymer aggregates: From micelles to vesicles and their biological applications. *Macromol Rapid Commun*. 2009;30(4–5):267–277.

23. Discher DE, Eisenberg A. Polymer vesicles. *Science*. 2002;297(5583):967–973.

24. Palivan CG, Goers R, Najer A, Zhang X, Car A, Meier W. Bioinspired polymer vesicles and membranes for biological and medical applications. *Chem Soc Rev*. 2016;45(2):377–411.

25. Kukula H, Schlaad H, Antonietti M, Förster S. The formation of polymer vesicles or "Peptosomes" by polybutadiene-b lock-poly (L-glutamate) s in dilute aqueous solution. *J Am Chem Soc*. 2002;124(8):1658–1663.

26. Cornelissen JJ, Fischer M, Sommerdijk NA, Nolte RJ. Helical superstructures from charged poly (styrene)-poly (isocyanodipeptide) block copolymers. *Science*. 1998;280(5368):1427–1430.

27. Rodríguez-Hernández J, Lecommandoux S. Reversible inside-out micellization of pH-responsive and water-soluble vesicles based on polypeptide diblock copolymers. *J Am Chem Soc*. 2005;127(7):2026–2027.

28. Holowka EP, Pochan DJ, Deming TJ. Charged polypeptide vesicles with controllable diameter. *J Am Chem Soc*. 2005;127(35):12423–12428.

29. Bellomo EG, Wyrsta MD, Pakstis L, Pochan DJ, Deming TJ. Stimuli-responsive polypeptide vesicles by conformation-specific assembly. *Nat Mater*. 2004;3(4):244–248.

30. Tanisaka H, Kizaka-Kondoh S, Makino A, Tanaka S, Hiraoka M, Kimura S. Near-infrared fluorescent labeled peptosome for application to cancer imaging. *Bioconjug Chem*. Jan 2008;19(1):109–117.

31. Bangham AD, Standish MM, Weissmann G. The action of steroids and streptolysin S on the permeability of phospholipid structures to cations. *J Mol Biol.* Aug 1, 1965;13(1):253–259.

32. MacKintosh FC, Janmey PA. Actin gels. *Curr Opin Solid State Mater Sci.* 1997;2(3):350–357.

33. Li S, Palmer AF. Effect of actin concentration on the structure of actin-containing liposomes. *Langmuir.* 2004;20(11):4629–4639.

34. Cortese JD, Schwab B, Frieden C, Elson EL. Actin polymerization induces a shape change in actin-containing vesicles. *Proc Natl Acad Sci USA.* Aug 1, 1989;86(15):5773–5777.

35. Wu KC-W, Yang C-Y, Cheng C-M. Using cell structures to develop functional nanomaterials and nanostructures—Case studies of actin filaments and microtubules. *Chem Commun.* 2014;50(32):4148.

36. Storm C, Pastore JJ, MacKintosh FC, Lubensky TC, Janmey PA. Nonlinear elasticity in biological gels. *Nature.* May 12, 2005;435(7039):191–194.

37. Xu J, Tseng Y, Wirtz D. Strain hardening of actin filament networks regulation by the dynamic cross-linking protein α-actinin. *J Biol Chem.* Nov 17, 2000;275(46):35886–35892.

38. Prost J, Jülicher F, Joanny JF. Active gel physics. *Nat Phys.* 2015;11(2):111–117.

39. Köhler S, Schaller V, Bausch AR. Structure formation in active networks. *Nat Mater.* June 2011;10(6):462–468.

40. Van den Heuvel MG, Dekker C. Motor proteins at work for nanotechnology. *Science.* 2007;317(5836):333–336.

41. Goel A, Vogel V. Harnessing biological motors to engineer systems for nanoscale transport and assembly. *Nat Nanotechnol.* 2008;3(8):465–475.

42. Patolsky F, Weizmann Y, Willner I. Actin-based metallic nanowires as bio-nanotransporters. *Nat Mater.* Oct 2004;3(10):692–695.

43. Beck R, Deek J, Jones JB, Safinya CR. Gel-expanded to gel-condensed transition in neurofilament networks revealed by direct force measurements. *Nat Mater.* Jan 2010;9(1):40–46.

44. Srinivasan N, Bhagawati M, Ananthanarayanan B, Kumar S. Stimuli-sensitive intrinsically disordered protein brushes. *Nat Commun.* 2014;5:5145.

45. Sano K-I, Kawamura R, Tominaga T, Nakagawa H, Oda N, Ijiro K et al. Thermoresponsive microtubule hydrogel with high hierarchical structure. *Biomacromolecules.* May 9, 2011;12(5):1409–1413.

46. Surrey T, Nédélec F, Leibler S, Karsenti E. Physical properties determining self-organization of motors and microtubules. *Science.* May 11, 2001;292(5519):1167–1171.

47. Sanchez T, Welch D, Nicastro D, Dogic Z. Cilia-like beating of active microtubule bundles. *Science*. July 22, 2011;333(6041):456–459.

48. Sanchez T, Chen DTN, DeCamp SJ, Heymann M, Dogic Z. Spontaneous motion in hierarchically assembled active matter. *Nature*. Nov 15, 2012;491(7424):431–434.

49. Um SH, Lee JB, Park N, Kwon SY, Umbach CC, Luo D. Enzyme-catalysed assembly of DNA hydrogel. *Nat Mater*. Oct 2006;5(10):797–801.

50. Winfree E, Liu F, Wenzler LA, Seeman NC. Design and self-assembly of two-dimensional DNA crystals. *Nature*. 1998;394(6693):539–544.

51. Rothemund PW. Folding DNA to create nanoscale shapes and patterns. *Nature*. 2006;440(7082):297–302.

52. Han D, Pal S, Nangreave J, Deng Z, Liu Y, Yan H. DNA origami with complex curvatures in three-dimensional space. *Science*. 2011;332(6027):342–346.

53. Andersen ES, Dong M, Nielsen MM, Jahn K, Subramani R, Mamdouh W et al. Self-assembly of a nanoscale DNA box with a controllable lid. *Nature*. May 7, 2009;459(7243):73–76.

54. Zhao Y-X, Shaw A, Zeng X, Benson E, Nyström AM, Högberg B. DNA origami delivery system for cancer therapy with tunable release properties. *ACS Nano*. 2012;6(10):8684–8691.

55. Jiang Q, Song C, Nangreave J, Liu X, Lin L, Qiu D et al. DNA origami as a carrier for circumvention of drug resistance. *J Am Chem Soc*. Aug 15, 2012;134(32):13396–13403.

56. Jones MR, Seeman NC, Mirkin CA. Programmable materials and the nature of the DNA bond. *Science*. 2015;347(6224):1260901.

57. Pinheiro AV, Han D, Shih WM, Yan H. Challenges and opportunities for structural DNA nanotechnology. *Nat Nanotechnol*. 2011;6(12):763–772.

58. Tanaka Y, Morishima K, Shimizu T, Kikuchi A, Yamato M, Okano T et al. An actuated pump on-chip powered by cultured cardiomyocytes. *Lab Chip*. 2006;6(3):362.

59. Choi E, Lee SQ, Kim TY, Chang H, Lee KJ, Park J. MEMS-based power generation system using contractile force generated by self-organized cardiomyocytes. *Sens Actuators B Chem*. Nov 26, 2010;151(1):291–296.

60. Park S-J, Gazzola M, Park KS, Park S, Di Santo V, Blevins EL et al. Phototactic guidance of a tissue-engineered soft-robotic ray. *Science*. 2016;353(6295):158–162.

6 Interactions of Carbon Nanostructures with Lipid Membranes

A Nano–Bio Interface

Mildred Quintana and Said Aranda

CONTENTS

6.1 INTRODUCTION

Carbon nanotubes (CNT) are considered alternative materials for the design of advanced drug and gene delivery vectors, biosensors, disease detection systems, cancer therapy agents, and imaging labels (Bianco et al. 2005). These nanodevices are manufactured at molecular level, improving the pharmacological profile and therapeutic properties of small molecular drugs and conventional detection systems. The ability of CNT to translocate cell membranes and their potentially high loading capacity made them effective transporters for shuttling and delivering peptides, proteins, nucleic acids, and molecular drugs into living cells (Prato et al. 2008). Likewise, the surface properties of CNT, namely, hydrophobicity, π–π staking, and surface topology from structural to atomic level, can be designed specifically to reach the desired advanced functionality (Micoli et al. 2013). Unfortunately, very little is known about the interactions that guide the internalization of CNT into cells, and consequently, the translocation mechanisms remain unclear.

The determination of the exact mechanisms of cellular uptake dependent on the physical and chemical properties of CNT is essential for the further development of biomedical devices and therapeutics for implantation or administration (Lacerda et al. 2007). Additionally, nanocarbons can also interact with biological molecules, producing potential damage to cells in vivo (Fischer and Chan 2007). It is well known that CNT interact with proteins, membranes, cells, DNA, and organelles, establishing bio–nano interfaces that depend on colloidal forces and dynamic biophysicochemical interactions (Nel et al. 2009). These interactions lead to diverse processes such as the formation of protein coronas, particle wrapping, intracellular uptake, and biocatalytic processes. Thus, the development of experimental techniques and predictive models to describe bio–nano interface processes are critical for the controlled design of innovative and safe functional nanomaterials.

Overall, functionalized CNT stand as the most promising nanovectors for biomedical and pharmaceutical applications due to the possibility of the formation of covalent bonds with the carbon skeleton producing chemically stable hybrids and the possibility for noncovalent functionalization with bioactive molecules, including therapeutic drugs, proteins, and oligonucleotides for their easy relief.

There are a number of publications dealing with the interaction of carbon nanostructures with lipids, particularly with lipid membranes. In this chapter, we summarize the recent published outcomes on the interaction of carbon nanostructures with lipids. Lipid membranes are the most simplified representation of cell membranes, the first barrier that CNT affront in their encounter with living cells. We expect that the assessment of the field might enable connections between experimental and theoretical work giving insights into the physical mechanisms than guide internalization of CNT into cells and other biological processes such as toxicity, signaling, and biodegradation. In the following sections, we first describe the CNT–lipid system, the experimental evidence of the interaction of CNT with lipids, then we discuss the recently published results on experimental and theoretical models, and finally summarize some perspectives and conclusions.

6.2 LIPID MEMBRANE INTERACTIONS WITH CNT

Cell plasma membranes are formed from lipids which ranges typically from 2 to 4 nm of length and less than 0.2 nm of diameter. Instead, CNT are long filaments that can be as long as several micrometers and with radius ranging from few nanometers to dozens of nanometers. Both systems interact in some form that CNT made the way to absorb, translocate, and/or internalize inside a model cell system, such as a lipid planar membrane or a vesicle.

CNT are formed by carbon atoms as a rolled graphene layer and this rolling tube can be done in many ways; there are many possible orientations of the hexagons on the nanotubes, but the basic shape is a cylinder. Carbon is the sixth element of the periodic table and each carbon atom has six electrons with $1s^2$, $2s^2$, and $2p^2$ atomic orbitals. The $2s^2$ and $2p^2$ orbitals are occupied by four electrons that are weakly bounded. Carbon nanotubes have two types of bonds due to sp^2 hybridization: the σ bonds and the π bonds. The first types of bonds are related with the hexagonal network and the later interact via π–π stacking interactions with another CNT or another affine atom or molecule. The π bonds are represented as p_z (z is perpendicular to the plane) orbitals of the carbon atoms. Actually, CNT are found in two forms: single-walled carbon nanotubes (SWCNT) and multiwalled carbon nanotubes (MWCNT) (consisting in a coaxial array of SWCNT separated from one another by approximately 0.34 nm). The electrical and mechanical properties of carbon nanotubes are quite remarkable: high Young's modulus and tensile strength can be metallic, semiconducting or semi-metallic wire, and the potential applications are enormous as long as they have high quality and can be produced at industrial levels. CNT can be grown by various methods, including arc discharge, laser ablation, and chemical vapor deposition.

6.3 ASSESSING TOXICITY

Perhaps toxicology is the most developed field in the study of bioapplications of CNT. The toxicity of CNT has been attributed to their unique physical and chemical properties, such as high surface area and high surface reactivity (Ravichandran et al. 2010). The surface reactivity is correlated with the ability of CNT to produced free radicals or reactive oxygen species (ROS) as super oxide radical anions and hydroxyl radicals via the activation of oxidative enzymatic pathways producing oxidative stress. The transition metal contaminates used as catalyst during the synthesis of CNT are considered the major source of oxidative stress. Additionally, direct oxidation is produced by CNT through the activation of the redox machinery of cell enhancing the generation of ROS.

6.3.1 Lipid Membrane

Biological membranes are natural barriers of the cell against external factors. Additionally, membrane molecules and soluble proteins are responsible for cell communication. Membranes activate the target cells by interacting with receptors that regulate such interactions. Cell–cell communication can also take place by exosomes, an intriguing yet unanswered mechanism. Exosome-like lipidic microvesicles are present in body fluids such as blood, urine, amniotic fluid ascites, and pleural effusions under healthy and disease conditions (Raposo and Stoorvogel 2013). We consider it extremely interesting to study the possible defense mechanism developed by lipidic membranes to prevent or minimize the toxic effects of the engineered CNT nanomaterials. One response of the membrane to external influences is lipid peroxidation.

6.3.2 Lipid Peroxidation

The generation of ROS is one of the most widespread responses of organisms to external influence. Activated at low concentration, ROS act as mediators in cell functions. However, at high concentration ROS may lead to cellular damage by lipids, proteins, enzymes, and nucleic acids oxidation. The main oxygenated molecules implicated in oxidative stress responses are the lipid peroxidation products. The lipid oxidation is mainly produced by enzymatic and nonenzymatic reactions generated by free radicals. For nonenzymatic reactions, the oxidation rates are mainly dictated by the number of double bonds in the lipids polyunsaturated fatty acid (PUFA) residues. These reactions are scarcely influenced by the nature of the polar heads. Instead, enzymatic oxygenations of free PUFA are the major source of extra- and intracellular biological regulators (Shvedova et al. 2012). Polyunsaturated phospholipids undergo an enzymatic attack by selective phospholipases with specific enzymatic oxygenation of free PUFA.

6.3.3 Biological Mechanism

There are a number of biological mechanisms proposed to explain CNT effects in biological systems. The most significant is the generation of ROS with the subsequent lipid peroxidation and development of inflammatory reactions.

However, additional mechanisms could play an important role in the adverse effects of CNT, such as stated by Boczkowski and Lanone (2012):

Oxidative stress is generated by the imbalance between oxidants production and antioxidants defenses. Common biomarkers of oxidative stress are malone dialdehyde (MDA), products of lipid peroxidation such as 4 hydroxynonenal (4-HNE), and protein carbonyls. The interaction with CNT causes a higher cellular inflammatory response accompanied by an increased secretion of two major inflammatory cytokenes; TNF and IL-6.

Inflammation is an early event after the exposure of CNT (6–24 h) with the recruitment of neutrophil-driven infiltration accompanied by the release of protein inflammatory cytokines, tumor necrosis factor (TNF) alpha, interleukin B and 6 (IL-B, IL-6), monocyte chemo-attractant protein (MCO)-1, or macrophage inflammatory protein (MIP)-2 or CXCL-2.

Genotoxic potential and mutagenic effects of CNT such as micronucleus induction, chromosome aberration and DNA damage could result as an excess of ROS and the surface properties of CNT. However, exists a controversy if CNT can play a role in genotoxic effects. Definitively, it is necessary to perform more research in order to understand CNT mechanism for genotoxic effects.

Protein corona refers to the interactions between components of the biological milieu and nanoparticles. These interactions are dictated by the structural composition of the xenobiotic and mediated by surface bound proteins and lipids. All nanoparticles in biological fluids are dynamically coated with a protein/ lipid mixture dependent on the size, shape, and surfaces properties of the nanoparticle. Protein corona is a major determinant of the localization and effects of nanomaterials *in vivo*. Protein corona in CNT could present favorable effects, for example, CNT coated with some proteins can increase their biocompatibility.

Degradation, biopersistence, and systemic translocation of CNT are important issues that remain poorly characterized and understood. Enzymatic degradation by horseradish peroxidase (HRP) (Micoli et al. 2014) and myeloperoxidase (MPO) can exist *in vivo* and appears as a mechanism responding inflammation. However, the relevance of this phenomenon on CNT biopersistence has not been investigated yet. For example, there is no evidence of translocation between organs.

6.4 EXPERIMENTAL EVIDENCE OF CNT–LIPID MEMBRANE INTERACTIONS

6.4.1 Lungs

As lungs are the primary route of entry for inhaled nanoparticles into the human body, lung toxicity is of specific concern. The adverse effect of MWCNT in rat lung epithelial (LE) cells caused a number of consequences on cultured cells such as dose- and time-dependent increase in the formation of free radicals, the accumulation of peroxidative products, the loss of cell viability, and antioxidant depletion, followed by apoptosis after 24 h triggered by the stimulation of the signaling pathway through an increase in the activity of caspase-3 and caspase-8 in rat LE cells (Ravichandran et al. 2009).

In the intrapharyngeal installation of SWCNT, the activation of heme oxygenase-1 (HO-1), a marker of oxidative stress, in lung, aorta, and heart tissue in HO-1 reporter transgenic mice was observed (Li et al. 2007). It was found that C57BL/6 mice exposed to SWCNT developed aortic mtDNA damage at 7, 28, and 60 days after exposure, accompanied by changes in aortic mitochondrial glutathione and protein carbonyl levels; these modification have been related to cardiovascular diseases.

It was demonstrated by mass spectrometry–based oxidative lipidomics analysis that all major phospholipid classes revealed highly selective patterns of pulmonary peroxidation after inhalation exposure of mice to SWCNT. No oxidized molecular species were found in the two most abundant phospholipid classes: phosphatidylcholine and phosphatidylethanoamine (Tyurina et al. 2011). Peroxidation products were identified in three relatively minor classes of anionic phospholipids, cardiolipin, phosphatidylserine, and phosphatidylinositol, whereby oxygenation of polyunsaturated fatty acid residues also showed selective substrate specificity, as shown in Figure 6.1. This nonrandom peroxidation coincided with the accumulation of apoptotic cells in the lungs.

This finding suggests the involvement of enzymatic mechanisms since nonspecific radical-scavenging antioxidants are not expected to be effective in

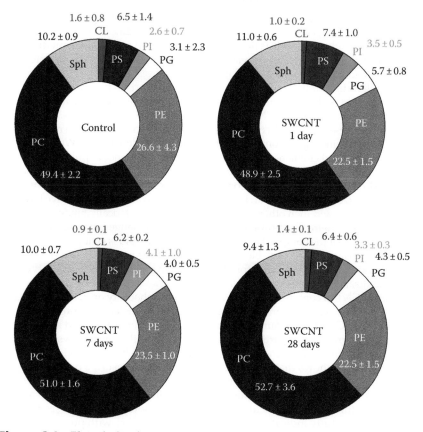

Figure 6.1 Phospholipid composition of the lung from control and SWCNT-treated mice on days 1, 7, and 28 after exposure. Lipids from lungs of control mice and mice exposed to SWCNT were isolated and resolved by 2D-HPTLC. Spots of phospholipids were visualized by iodine vapors. After that, spots were scraped and lipid phosphorus was detected. The content of phospholipids was expressed as percent of total phospholipids. (Reprinted with permission from Tyurina, Y.Y., Kisin, E.R., Murray, A., Tyurin, V.A., Kapralova, V.I., Sparvero, L.J., Amoscato, A.A. et al., Global phospholipidomics analysis reveals selective pulmonary peroxidation profiles upon inhalation of single-walled carbon nanotubes, *ACS Nano*, 5(9), 7342–7353. Copyright 2011 American Chemical Society.)

suppressing enzymatically driven lipid peroxidation. Then their application as protectants against SWCNT-induced pulmonary damage is questionable. Exposure to SWCNT triggers specific pathways of cellular damage, which implies that the adverse effects caused by SWCNT could be counteracted by targeting these pathways, likely apoptotic in nature. These observations suggest the involvement of mitochondria-dependent apoptosis as well as macrophage disposal of apoptotic cells in the regulation of the inflammatory response to SWCNT.

The adverse effects of CNT in the pulmonary system can be reduced by treating CNT before exposure or by the use of preventive treatments. Different studies have shown that antioxidant treatments, both endogenous and dietary, can protect organs from damage by oxidative stress. The exposure of mice to SWCNT induces unusually robust pulmonary response with an early onset of fibrosis, accompanied by oxidative stress and antioxidant depletion. C57BL/6 mice maintained in a Vitamin E (VE)–deficient diet, a major lipid-soluble antioxidant, caused a 90-fold depletion of alpha-tocopherol in the lung tissue and resulted in a significant decline of other antioxidants such as GSH, ascorbate, as well as the accumulation of lipid peroxidation products (Shvedova et al. 2007). A greater decrease of pulmonary antioxidants was detected in SWCNT-treated VE-deficient mice as compared to controls.

A very interesting strategy preventing CNT damage starts in the alveoli, since CNT first interact with the pulmonary surfactant. At this interface, proteins and lipids of the surfactant bind to CNT changing their surface characteristics. The effects caused by the presence of CNT covered with surfactant biomolecules are not well understood. The binding of biomolecules to different functionalized MWCNT and their entry into blood circulation by inhalation, with the presence of pulmonary surfactant coating on the CNT surface, demonstrated that lipids and proteins bind to MWCNT changing their surface properties (Gasser et al. 2010). Proteins of the pulmonary surfactant bind to CNT; bound lipids seem to enable the selective binding of plasma proteins and then other plasma proteins may be sterically hindered to bind CNT. Like proteins, lipids also undergo dynamic exchange processes. There is an indication that the composition of bound surfactant lipid is changed when MWCNT come into contact with blood plasma lipids. The functionalization of MWCNT was identified as a factor for plasma protein binding. However, the type of functionalization, amino or carboxylic group, causes a minor role on the surface properties of CNT in contrast to the alteration in hydrophobicity or steric hindrance that resulted from surfactant and protein binding. The biomolecules absorbed to the surface of CNT trigger numerous functions, for example, transport and uptake mechanisms of nanoparticles or fulfill functions in the immune system. An uncertainty remains as new functions can be expected from bound proteins.

A similar study shows that MWCNT wrapped with Curosurf, a pulmonary surfactant, affects their oxidative potential by increasing the generation of ROS and decreasing intracellular glutathione depletion in MDM as well as decreases the release of tumor necrosis factor alpha (TNF-α), followed by apoptosis (Gasser et al. 2012). This study indicates that precoating of MWCNT with pulmonary surfactant more than functionalization of CNT is a key factor in determining their ability to cause oxidative stress, cytokine/chemokine release, and apoptosis; a schematic representation is shown in Figure 6.2.

In order to understand the importance of the adsorption of lung surfactant on CNT, an established mouse model of pharyngeal aspiration of SWCNT was used (Kapralov et al. 2012). In this work, the aspirated SWCNT were recovered from the bronchoalveolar lavage fluid (BALf). Then, the recovered SWCNT were purified from possible contamination with cells and the composition of phospholipids was evaluated by liquid chromatography mass spectroscopy (LC-MS). It was found that SWCNT selectively absorbed two types of surfactant

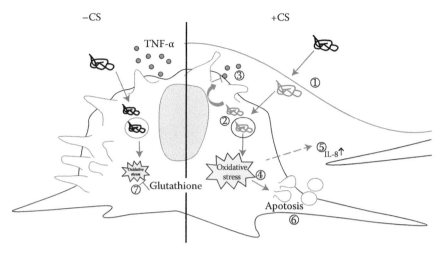

Figure 6.2   Scheme of observed effects from MWCNT, coated with Curosurf (+CS) or without (−CS) and the proposed underlying mechanism. Lipids and proteins of the surfactant bind to the MWNCTs and alter their surface characteristics (1). Surfactant-coated tubes are subsequently located in vesicles of MDM and free in the cytoplasm (2). After 24 h a decrease in TNF-α release (3) is observed, which might be due to a downregulation of TNF-α mRNA by Curosurf compounds. And increase in ROS (4) causes further an increase in IL-8 chemokine release in epithelial cells (5) and induction of apoptosis (6) in MDM. Uncoated MWCNT, which are present inside MDM after 24 h exposure, are inducing and intracellular gluthione depletion (7). (Adapted with permission from Gasser, M., Wick, P., Clift, M.J.D., Blank, F., Diener, L., Yan, B., Gehr, P., Krug, H.F., and Rothen-Rutishauser, B., Pulmonary surfactant coating of multi-walled carbon nanotubes (MWCNT) influences their oxidative and pro-inflammatory potential in vitro, *Particle and Fibre Toxicology*, 9(1), 17. Copyright 2012 BioMEd Central.)

phospholipids: phosphatidylcholines (PC) and phosphatidylglycerols (PG). The surfactant lipids, uninterrupted, coated the SWCNT surface with the polar head groups pointed away from the carbon skeleton. The presence of surfactant proteins (a), (b), and (d) on SWCNT was also found, as shown in Figure 6.3.

6.4.2 Liver

Nonalcoholic fatty liver disease (NAFLD) is characterized by the accumulation of lipids in the liver. The disease starts as a relatively benign condition as a simple steatosis that unfortunately progresses into steatohepatitis, which is a more severe stage of NAFLD. In this condition, lipids accumulation occurs with inflammation. Exposure to carbon particles has been suggested to generate hepatosteatosis by an oxidative stress mechanism. As we have already described, oxidative stress and inflammation are key mechanisms of action of particle-generated health effects. Particles can generate ROS and decrease the activity of antioxidant defense system.

In this direction, the surface functionalization of carbon nanomaterials plays an important role in their interaction within biological systems. Mice were exposed to 10 mg/kg of pristine and acid-oxidized MWCNT with variable degree of carboxylation by intravenous injection (Jain et al. 2011). After that, short-term (7 days) and long-term (28 days) impact to MWCNT were evaluated and found to induce significant hepatotoxicity and oxidative damage in mice. However,

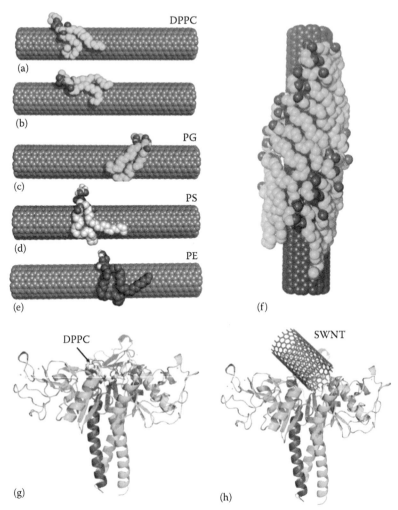

Figure 6.3 Computer modeling of SWCNT binding with phospholipids and SO-D. The predictive binding pose of (a) the lowest energy conformation of DPPC, (b) DPPC bound along the axis of SWCNT, the lowest energy conformation of (c) PG, (d) PS, and (e) PE. (f) Lipid coating model generated using the PG-bound form of SWCNT is represented as spheres and colored in gray. The different phospholipids DPPC, PG, PS, and PE are rendered as spheres and colored in cyan, green, white, and magenta, respectively. In all cases, the N, O, and P atoms are colored in blue, red, and orange, respectively. The predictive binding sites of DPPC (g) and SWCNT (h) are on SP-D. The 3D structure of SP-D is colored according to the different chains and is represented as a cartoon. The structures of both DPPC and SWCNT are represented as sticks. The SWCNT is colored in gray, and the structure of DPPC is colored based on its atoms, that is, carbon, oxygen, and nitrogen atoms in yellow, red, and blue, respectively. (Reprinted with permission from Kapralov, A.A., Feng, W.H., Amoscato, A.A., Yanamala, N., Balasubramanian, K., Winnica, D.E., Kisin, E.R. et al., Adsorption of surfactant lipids by single-walled carbon nanotubes in mouse lung upon pharyngeal aspiration, *ACS Nano*, 6(5), 4147–4156. Copyright 2012 American Chemical Society.)

the damage was recovered after 28 days of treatment without MWCNT exposure. Overall, oxidized MWCNT were less toxic and more biocompatible than their pristine counterparts. Toxicological parameters suggested that toxicity of MWCNT critically depends on their functionalization density.

A different study using SWCNT caused significant damage to DNA and cytotoxicity on human hepatocarcinoma cells (HepG2) in a dose- and time-dependent manner (Alarifi et al. 2014). The ROS generation and lipid peroxidation are the main factors in the induced cellular damage. In this work, DNA fragmentation analysis showed that SWCNT caused genotoxicity.

6.4.3 Nervous System

Studies have demonstrated that ultrafine carbon particles are able to cross the blood–brain barrier and impact on the central nervous system; here the necessity is to understand the possible interaction between CNT and neural cells (Oberdörster et al. 2005).

In this direction, the concentration-dependent cytotoxicity of SWCNT and SWCNT functionalized with polyethylene glycol (SWCNT-PEGs) in neural PC12 cells at the biochemical, cellular, and gene expressional levels, as illustrated in Figure 6.4, showed that SWCNT-PEGs significantly reduce the side effects of SWCNT in vitro in neuronal PC12 cells as well as the disturbance of oxidative stress-related gene expression (Zhang et al. 2011).

Given their relatively lower induced toxicity, the surface functionalization of CNT is expected to be a promising strategy for delivering drugs and growth factors to neuronal cells, with the potential in treating major neurodegenerative diseases. Additionally, antioxidant treatments, both endogenous and dietary, can protect nervous tissue from damage by oxidative stress (Contestabile 2001). SWCNT can induce

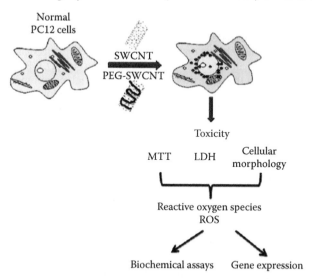

Figure 6.4 Diagram of the experimental procedure. SWCNT and SWCNT-PEGs were incubated with PC12 cells, and their toxicity was assessed by a combination of biochemical and gene expression approaches. (Reprinted with permission from Zhang, Y., Xu, Y., Li, Z., Chen, T., Lantz, S.M., Howard, P.C., Paule, M.G. et al., Mechanistic toxicity evaluation of uncoated and PEGylated single-walled carbon nanotubes in neuronal PC12 cells, *ACS Nano*, 5(9), 7020–7033. Copyright 2011 American Chemical Society.)

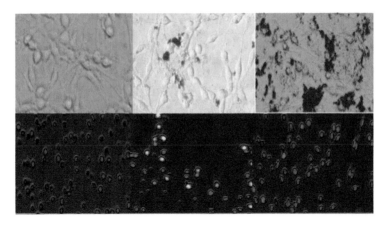

Figure 6.5 Observation of SWCNT-induced PC12 cell morphological changes after treatment with VE PC12 cells were pretreated with 1 mM VE for 1 h prior to SWCNT (50 μg/mL) for 48 h. The cells were observed through an optic microscope directly or fluorescence microscopy (200×) after nuclei staining with Hochest 33258. (Reprinted from *Toxicology in Vitro: An International Journal Published in Association with BIBRA*, 26(1), Wang, J., Sun, P., Bao, Y., Dou, B., Song, D., and Li, Y., Vitamin E renders protection to PC12 cells against oxidative damage and apoptosis induced by single-walled carbon nanotubes, 32–41, Copyright 2012, with permission from Elsevier.)

apoptosis and oxidative damage on PC12 cells, an in vitro model of neuronal cells. It was shown that the neuroprotective effects of vitamin E (VE) on SWCNT-induced neurotoxicity in cultured PX12 cells (Wang et al. 2012). Authors demonstrated that VE increased PC12 cells viability and significantly attenuated SWCNT-induced apoptotic cell death in a time- and dose-dependent manner, Figure 6.5.

The mechanism is mainly based on the reduction of ROS, the decreased level of lipid peroxide, the elevated level of GSH, and activities of SOD, GPx, and CAT. Additionally, VE blocked the reduction in the mitochondrial membrane potential and the activation of the caspase-3. Basically, biochemical and morphological studies demonstrated VE protect PC12 cells from the injury induced by SWCNT by the regulation of oxidative stress and prevention of mitochondrial-mediated apoptosis.

6.4.4 Blood

Macrophages are the primary responders to different external particles initiating and propagating inflammatory reactions and oxidative stress. The interaction of RAW 264.7 macrophages and two types of SWCNT, iron-rich (26 wt.% of iron) and iron-stripped (0.23 wt.% of iron) SWCNT, showed that neither iron-rich nor iron-stripped SWCNT were able to generate intracellular production of super-oxide radicals or nitric oxide (Kagan et al. 2006). Different SWCNT displayed different redox activity in a cell-free model system as revealed by EPR-detectable formation of ascorbate radicals resulting from ascorbate oxidation. Iron-rich SWCNT caused a significant loss of intracellular low molecular weight thiols (GSH) and the accumulation of lipid hydroperoxides in RAW 264.7 macrophages. From this study, it was concluded that the presence of iron in SWCNT may be important in determining redox-dependent responses of macrophages.

In a different experiment, exposure of MWCNT to rats produced a significant dose-dependent reduction of blood total antioxidant capacity, glutathione, superoxide dismutase, catalase activity, and increased malondialdehyde levels (Reddy et al. 2011). The proposed mechanism behind the MWCNT-induced oxidative stress and cytotoxicity may be due to the ability of MWCNT preferentially mobilized to mitochondria. Since mitochondria are redox active organelles, there is a likelihood of altering ROS production and thereby overloading or interfering with antioxidant defenses.

Similar observations were found in human lymphocytes, MWCNT containing in their composition metal nanoparticles, that induce the excessive formation of ROS and consequentially oxidative stress was present in the cells. In this study, the generation of ROS was attributed to the presence of iron nanoparticles since the addition of an iron chelator, deferoxamine, during the experiments, considerably reduces the generation of ROS (Zhornik et al. 2012).

6.4.5 Muscle

CNT are promising materials for the development of biomedical scaffolds with excellent properties, such as conductivity and strength. Then, understanding the interaction of CNT with muscular tissue is paramount. MWCNT modulate the proliferation and differentiation of skeletal muscle cell line C2C12 (Tsukahara and Haniu 2011). MWCNT stimulated intracellular lipid accumulation in C2C12 cells. When MWCNT were added to C2C12 myocytes, they incorporated into the cells and promoted their differentiation into adipose-like cells. Cells maintained in culture medium clearly differentiated into myotubes, while exposure to MWCNT significantly reduced the expression of myogenin and led to the emergence of adipocyte-related gene markers. The expression of adipose-related genes was markedly upregulated during MWCNT exposure. In this work it was concluded that the exposure to MWCNT converts the differentiation pathway of C2C12 myoblast into that of adipoblast-like cells, as shown in Figure 6.6. This study suggested that MWCNT significantly inhibits myotube formation and the expression of muscle specific genes.

6.4.6 Kidney

The cytotoxicity of two different sizes of MWCNT in cultured human embryonic kidney (HEK293) cells found that the exposure to MWCNT caused dose-depended cytotoxicity as revealed by the mitochondrial function and cell viability measured by the MTT assay (Reddy et al. 2010). The cell membrane damage induced by MWCNT was monitored by the LDH leakage assay. LDH is a stable cytosolic enzyme in normal cells that leaks into the extracellular fluid only after membrane damage. MWCNT increase LDH leakage in a dose-dependent manner at 48 h exposure period. Exposure to MWCNT for 48 h caused a significant time-dependent increase in the concentrations of the proinflammatory cytokine interleukin-8 (IL-8) released from HEK293 cells, indicating an inflammation response to MWCNT. Concomitant cellular oxidative stress was manifested by reduced cellular levels of glutathione and increased lipid peroxidation, followed by the production of malondialdehyde. Lactate dehydrogenase leakage from cells is considered another evidence for penetration into the membrane and cell membrane damage. These observations indicate that increased lipid peroxidation was the primary cause of membrane damage and cytotoxicity in HEK kidney cells.

6.4.7 Intestine

Carboxylic acid functionalized (COOH-SWCNT) reduce cell viability and induce morphological effects on the human intestinal line Caco-2 (Jos et al. 2009). The proposed mechanism involves oxidative stress response of this line against COOH-SWCNT, accompanied by an increase in ROS generation, enzymatic and nonenzymatic antioxidant defenses. However, the antioxidant defenses could not overwhelm the oxidative aggression caused by COOH-SWCNT, a lipid peroxidation product increased in a concentration-dependent manner.

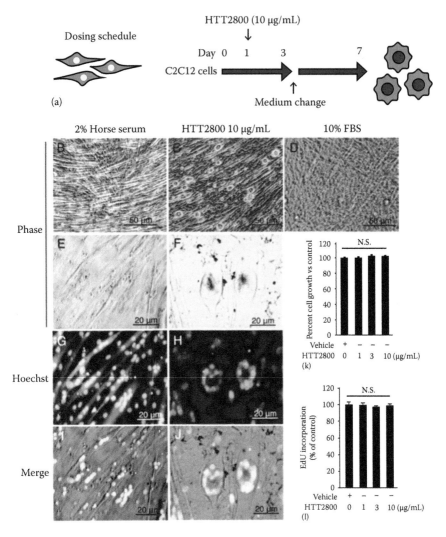

Figure 6.6 Effect of HTT2800 on the morphology of C2C12 skeletal muscle cells. (a) Schematic showing the protocol for HTT2800-induced cell morphology. Cells were cultured in growth medium until 80%–90% confluency and then in medium with or without HTT2800 (10 μg/mL). (b) Seven days of C2C12 cell differentiation under a low-serum condition (2% horse serum) led to changes in morphology from myoblasts to myotubes as observed by light microscopy. (c and d) Cells were maintained to day 7 after reaching confluence in DMEM with 10% FBS without additions or in the presence of 10 μg/mL HTT2800. Cellular morphology was visualized on day 7 by using light microscope. (g and h) Cell nuclei stained with Hoechst 33342 were visualized on day 7 by using a fluorescent microscope. (i and j) Merge image of phase contrast and fluorescence. (Reprinted from *Biochemical and Biophysical Research Communications*, 406(4), Tsukahara, T. and Haniu, H., Nanoparticle-mediated intracellular lipid accumulation during C2C12 cell differentiation, 558–563, Copyright 2011, with permission from Elsevier.)

6.4.8 Skin

The main function of the skin is to protect the organism from the external chemical and biological environment. Some studies had been performed in order to understand the protective behavior of skin against carbon nanoparticles, for example, the penetration of fullerene-FITC-peptide of approx. 3.5 nm dispersed in PBS was studied using dermatomed porcine skin, flow through diffusion cells, and receptor fluid perfused at 2 mL/h (Baroli 2010). Results showed that fullerenes diffused into the skin by passive penetration as a function of flexing time and exposure.

6.4.9 Other Cells

6.4.9.1 Stem Cells

By fluorescence recovery, after the photobleaching (FRAP) technique, it was found that CNT affects the membrane diffusion of cells. In this experiment, human mesenchymal stem cells were cultured on aligned SWNT networks (Park et al. 2014). The molecules in the cell membrane were found to diffuse faster along the direction parallel to the aligned CNT than along the direction orthogonal to the network. This study demonstrated that the nanoscale properties of nanostructured materials may significantly affect molecular diffusion in cell membranes and could be related with other cellular processes such as cell adhesion and cellular signaling, a schematic representation is shown in Figure 6.7.

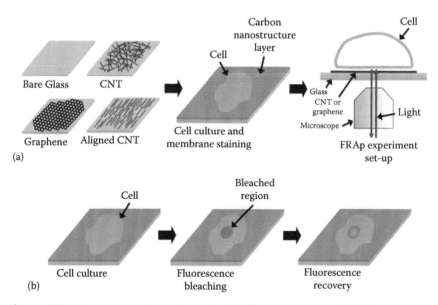

Figure 6.7 Experimental procedure for FRAP experiments on hMSCs cultured on carbon nanostructured substrates. (a) Schematic diagram depicting experimental procedure to prepare hMSCs on glass, CNT, and graphene substrates. Bare glass, CNT, and graphene substrates were prepared, and hMSCs were seeded on the substrates for FRAP experiments. The substrates were mounted on an inverted microscope equipped with a mercury lamp and an electron multiplying charged coupled device (EMCCD). (b) Schematic diagram depicting the procedure of FRAP experiments. The specific region of an hMSC was photobleached, and the fluorescence images were obtained every 30 s. (Reprinted with permission from *Journal of Physical Chemistry C*, 118(7), 3742–3749. Copyright 2014 American Chemical Society.)

6.4.9.2 Breast Cancer Cells

For the functionalization of SWCNT with Palitaxel (PTX), a drug for breast cancer, a nontoxic lipid molecule docosanol was conjugated first with SWCNT. Folic acid was also conjugated with SWCNT for targeted drug delivery achieving high loading onto SWCNT and high cell penetration capacity, resulting in improved drug efficacy in vitro in comparison to free drug Taxol (Shao et al. 2013). In vivo analysis, using a human breast cancer xenograft in mice model also confirmed the improved drug efficacy of the nanovector, see Figure 6.8.

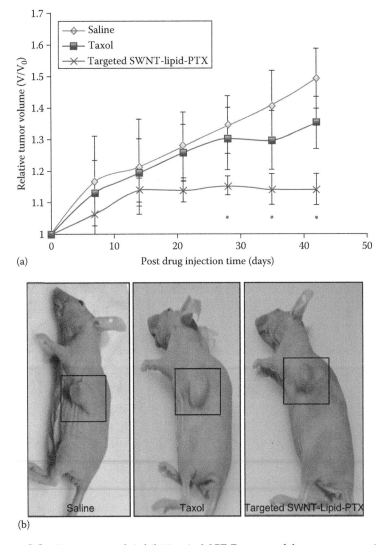

(a)

(b)

Figure 6.8 Tumor growth inhibition in MCF-7 xenograft breast cancer mice model. (Reprinted from *Biomaterials*, 34(38), Shao, W., Paul, A., Zhao, B., Lee, C., Rodes, L., and Prakash, S., Carbon nanotube lipid drug approach for targeted delivery of a chemotherapy drug in a human breast cancer xenograft animal model, 10109–10119, Copyright 2013, with permission from Elsevier.)

The targeted SWCNT-lipid PTX was found nontoxic, as evaluated by biochemical analysis using blood samples and histological analysis of the major organs.

6.4.9.3 Maternal Exposure

The potential effects of MWCNT on pregnant dams and embryo-fetal development in rats surprisingly showed that the only effect of MWCNT was a decrease in thymus weight. Maternal body weight, food consumption, and oxidant–antioxidant balance in the liver, gestation index, fetal deaths, fetal and placenta weights, or sex ratio were not affected by the treatment. Importantly, in this experiment, all animals survived till the end of the study (Lim et al. 2011).

6.4.10 Daphnia magna

Grazing zooplankton (*Daphnia magna*) could ingest lipid-coated SWCNT from the water during their normal feeding. Acute mortality was observed at high concentrations (>5 mg/L). However, when natural organic matter was used in place of the lipid to stabilize CNT dispersion, the toxicity of CNT was argued to have been feeding inhibition, leading to a deficit in nutrient intake (Alloy and Roberts 2011).

Additionally, *D. magna* could modify the solubility of SWCNT coated with lysophosphatidylcholine likely through the digestion of the lipid coating, see Figure 6.9. This study provides the evidence of biomodification of CNT material by an aquatic organism. Acute toxicity was observed only in the highest concentration test (Roberts et al. 2007).

6.4.11 Bacteria

Purple membranes composed of bacteriorhodopsin (bR) and their native surrounding lipids interacting with pristine SWCNT in a buffer solution form small bundles with an average thickness of 1.5 nm, as indicated in Figure 6.10 (Bertoncini and Chauvet 2010). Membrane bR proteins stabilized SWCNT in buffer dispersion through hydrophobic interactions between the bR alpha-helices and the sidewall of CNT.

Later, it was also observed that different types of CNT influence the viability, composition of fatty acids, and cytoplasmic membrane fluidity of three different bacteria in aqueous medium (Zhu et al. 2014). The cytoplasmic membrane fluidity of bacteria increased with CNT concentration, and a significant negative correlation existed between the bacterial viabilities and membrane fluidity for *Escherichia coli* and *Ochrobactrum*. This result indicates that the increase in membrane fluidity induced by CNT is an important factor in the inactivation of bacteria. An elevation

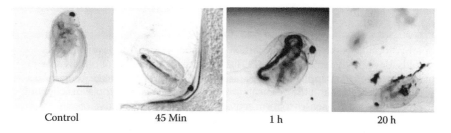

Control   45 Min   1 h   20 h

Figure 6.9 Time course micrographs of *D. magna* exposed to 5 mg/L of lyso-phosphatidylcholine-coated SWCNT. (Reprinted with permission from Roberts, A.P., Mount, A.S., Seda, B., Souther, J., Qiao, R., Lin, S., Pu, C.K., Rao, A.M., and Klaine, S.J., In vivo biomodification of lipid-coated carbon nanotubes by *Daphnia magna*, *Environmental Science and Technology*, 41(8), 3028–3029. Copyright 2007 American Chemical Society.)

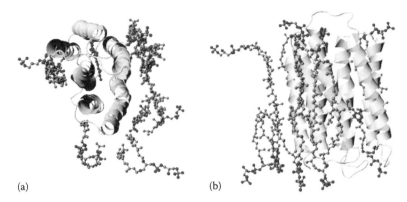

(a)   (b)

Figure 6.10 Top view (a) and side view (b) of the schematic representation of the 3D structure of bR in a purple membrane (i.e., with its surrounding lipids chains). The seven alpha-helical bundles from a transmembrane pore. The retinylidene residue (in blue) is linked to the protein moiety via a Schiff base linkage to lysine-216. (Reprinted with permission from Bertoncini, P. and Chauvet, O., Conformational structural changes of bacteriorhodopsin adsorbed onto single-walled carbon nanotubes, *Journal of Physical Chemistry B*, 114(12), 4345–4350. Copyright 2010 American Chemical Society.)

in the level of saturated fatty acids accompanied with reduction in unsaturated fatty acids, compensates for the fluidizing effect of CNT and demonstrates that the bacteria are able to modify their composition of fatty acids to adapt to the toxicity of CNT. For, *Staphylococcus aureus* and *Bacillus subtilis* exposed to CNT increased the portion of branched-chain fatty acids and decreased the level of straight-chain fatty acids, and this was favorable to counteract the toxic effect of CNT. Bacteria tolerances to CNT are associated with both the adaptive modification of fatty acids in the membrane and the physicochemical properties of CNT.

Additionally, the effects of long and short MWCNT were evaluated in gram-negative bacteria. Attenuated total reflection Fourier transform infrared (ATR-FTIR) spectroscopy was used to find a concentration-dependent response in the spectral alteration to lipids, amide II, and DNA (Riding et al. 2012). The observed biomolecular alterations following exposure to MWCNT did not totally inhibit cellular metabolism, since the toxicity of CNT was insufficient to result in the death of the entire cell population.

6.4.12 Protozoa

Superhydrophilic vertically aligned carbon nanotubes (VACNT-O_2) were used as scaffolds for photodynamic therapy (PDT) to induce the inhibition of cell division in eukaryotic cells.[39] PDT applied to the parasitic protozoan *Tritrichomonas foetus* induced the generation of ROS and the permeabilization of lysosomes in cytoplasm activating the processing of procaspases to form active caspase. Finally, the active caspase–induced apoptosis. This mechanism induced lysosome plasma membrane fusion, releasing the content into the extracellular microenvironment. The release of protease leads to the destruction of the basal lamina favoring the process pathogenesis.

6.5 COMPUTER SIMULATIONS

The whole cells have a complex internal structure, and the external membrane, mainly composed of proteins, lipids, and cholesterol, offers a boundary to protect the internal contents. The experimental methods described in Section 6.4 describe the different aspects of the CNT interaction with this membrane. In order to gain more insight at molecular level, we need to use powerful techniques of the computer simulations. Several simulations studies have been reported to explore CNT and lipid membranes. The formulation of a model describing the interaction between CNT and lipids membranes is challenging. The CNT wrapping by lipids involves weak bonds such as van der Waals interactions and London forces that depend on the CNT properties as chirality, diameter, purity, etc. Three main aspects of CNT simulations have been performed: lipid bilayer deformation, penetration, and artificial channel membranes by CNT. An excellent review of this work was done by Monticelli et al. (2009) where the main conclusions are the following: (1) In the case of CNT as artificial membrane channels, the hydrophobic mismatch between the CNT and lipids and the interactions of the lipid head groups and CNT functional groups affect the efficiency of CNT as transmembrane channels. (2) Molecular simulation studies show that CNT have an effect on the local structure and dynamics of the membrane. (3) Diffusion of lipids on the membrane containing CNT is decreased in proportion to the CNT diameter. (4) The effect of the hydrophobic mismatch at the nanotube-lipid boundary was studied and they found three different effects on the lipid bilayer thickness. No effects were observed on the thickness by nanotubes far away from the lipid bilayer. (5) By distances similar to the lipid length the lipid bilayer is thicker, and by closer distances the lipid bilayer thins or thickens with shorter or longer nanotubes, respectively. (6) In the case of CNT permeation through the membrane, SWNTs and MWNTs were functionalized, making strong attraction to the head group of the lipid, and they found that MWNTs were wrapped by the lipid head groups, enfolding it and leaving a narrow neck on the other side, and by decreasing the strength of the interaction in the neck region, a fission of a small vesicle containing the CNT could be induced. (7) For small SWCNT (diameter 1.0 nm) the nanotube simply pierced inside the bilayer and remained there with its long axis parallel to the bilayer normal. (8) Functionalized CNT with hydrophobic structures could be inserted in cell membranes. Pristine SWCNT with diameters between 1.4 and 6.1 nm were pulled through a lipid membrane and the CNT internalization was associated with bilayer deformations. For more details refer Monticelli et al. (2009) and the works cited in it. We will focus on the aggregation and internalization of CNT.

6.5.1 Aggregates

CNT can form different types of aggregates in solutions and idealized simulation results can be very difficult for experimental verification. Different simulation methods have been implemented to elucidate the main parameters yielding the equilibrium configurations of CNT interacting with different types of solvents and lipids. Classical molecular dynamics simulations have been tested for atomistic models of nanoparticles near a lipid bilayer membrane and have been found that particle morphology has effects on the structure, mobility, and adsorption of these particles. The carbon nanoparticles prefer to stay in the hydrocarbon tail regions of the lipids and move freely with speed depending on their molecular weight (Chang and Violi 2006). The study of water transport through nanostructures has been studied by the simulation method; for example, using the continuous-time random walk model water fills a (6,6) CNT spontaneously (Hummer et al. 2001; Berezhkovskii and Hummer 2002) and a unidirectional diffusion of water within CNT across a lipid membrane under equilibrium conditions and under the influence of static and alternating fields and this diffusion is reduced with the increase of the electric field frequency (Garate et al. 2009a,b). Molecular dynamics simulations reveal that interactions between CNT and Lipids (DPPC) is concentration dependent (Wang et al. 2009) where the hydrophobic part of the lipid is absorbed

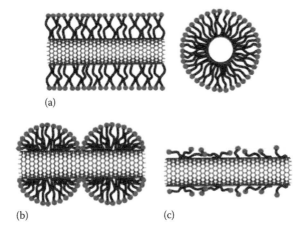

(a)

(b)   (c)

Figure 6.11   Illustration of the adsorption models for CNT–Lipid interaction: (a) cylindrical micelle, (b) hemimicelle, and (c) random adsorption. (Adapted with permission from Wallace, E.J. and Sansom, M.S.P., Carbon nanotube self-assembly with lipids and detergent: A molecular dynamics study, *Nanotechnology*, 20(4), 045101. Copyright 2009 IOP.)

on the surface of the CNT and the hydrophilic part is oriented toward the aqueous phase. A coarse-grained molecular dynamics shows that by increasing the lipid concentration, DPPC Lipids encapsulate the CNT within a cylindrical micelle (Wallace and Sansom 2009). In the last study they illustrate the possible adsorption scenarios: cylindrical micelle, hemimicelle, and random adsorption, see Figure 6.11, suggesting that the adsorption mechanism is concentration dependent. Using the same simulation method shows that hydrated lipid micelles filled with hydrophobic molecules can be self-assembled on the surfaces of CNT (Patra and Král 2011).

The phospholipid membranes of the cells are designed to protect the inner part and in function as a communication transmitter of the external changes in the medium. We discuss in Section 6.3 the different aspects of CNT internalization and the effects of the presence of CNT on different kind of cells. To study this effect we need to answer the question: Can CNT penetrate a lipid bilayer? And, how do the properties of CNT influence this internalization? Experimentally, it is difficult to visualize the interaction of a simple CNT with a lipid bilayer (even a bundle of CNT) and even if it is possible, it will be a challenge to access the time and space scale of a CNT–lipid bilayer interaction. This is another reason why simulation methods are important and are necessary to elucidate the main aspects of this encounter.

6.5.2 Internalization

CNT have a great efficiency to penetrate the living cells, as we discussed in Section 6.3, but the translocation mechanisms are not clear yet. Again, instead of discussing the piercing of CNT on the cell, we focus on model lipid membranes. Two main translocation mechanisms have been proposed: one dependent energy that relates endocytosis processes and one independent energy that relates free diffusion through the lipid bilayer. The spontaneous piercing of the lipid membrane driven only by thermal fluctuations was studied using the single-chain mean field theory and authors found that the energy cost of bilayer rupture is high compared to that of the energy thermal motion, supporting the theory of the energy-dependent translocation mechanism (Pogodin and Baulin 2010).

Insertion of CNT and CNT bundles using two different techniques is done as follows: First a free energy landscape is calculated based on different insertion geometries; then the dynamics is investigated using coarse-grained approach to study the interaction process. They found that CNT prefers horizontal orientation inside the internal hydrophobic layer of the membrane (Höfinger et al. 2011). The dependence on the chirality and length of SWCNT in the internalization of lipid membranes was studied using molecular dynamics simulations (Skandani et al. 2012). In this case, authors found that CNT with higher aspect ratios internalize the membrane faster while shorter nanotubes undergo rotations during the final stage of endocytosis, see Figure 6.12. Another interesting result was that CNT with lower chiral indices develop adhesion with the membrane.

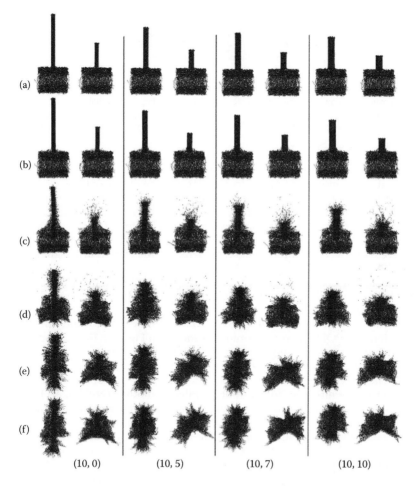

Figure 6.12 Sequential representation of SWCNT penetration through the cell membrane: (a) initial configuration, (b) after 0.4 ps, (c) after 4 ps, (d) after 10 ps, (e) after 30 ps, and (f) after 60 ps. (Reprinted with permission from Skandani, A.A., Zeineldin, R., and Al-Haik, M., Effect of chirality and length on the penetrability of single-walled carbon nanotubes into lipid bilayer cell membranes, *Langmuir: The ACS Journal of Surfaces and Colloids*, 28(20), 7872–7879. Copyright 2012 American Chemical Society.)

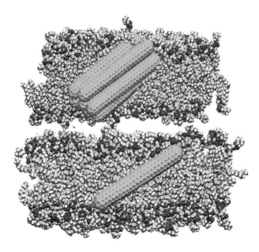

Figure 6.13 Representative simulation snapshots of the systems. (Reprinted with permission from Parthasarathi, R., Tummala, N.R., and Striolo, A., Embedded single-walled carbon nanotubes locally perturb DOPC phospholipid bilayers, *Journal of Physical Chemistry B*, 116(42), 12769–12782. Copyright 2012 American Chemical Society.)

Using all-atom molecular dynamics simulations, the effects of CNT embedded within DOPC lipid membrane were studied. The lipid membrane structure, organization, and dynamic perturbation are very short ranged, but it is larger for a bundle of CNT. The presence of CNT is found to reduce the lipid mobility within the membrane and to perturb the structure of interfacial water, see Figure 6.13, of the systems considered in this work (Parthasarathi et al. 2012).

Functionalized CNT passively penetrate the lipid membrane; this was demonstrated by molecular dynamics methods combined with Monte Carlo simulations (Kraszewski et al. 2012).

Triethylene glycol–functionalized SWCNT were simulated to study the penetration in a POPC lipid membrane, as shown in Figure 6.14. Authors propose that the main driving force in the uptake mechanism and surface coating asymmetry in the cell membrane could strongly influence the CNT uptake. The effects of embedded SWCNT, double-walled carbon nanotubes (DWCNT), or the nitrogen-doped double-walled nanotube (N-DWCNT) into DMPC lipid membranes using molecular dynamics simulations found that the steric interactions of this system lead to the reduction of the entropy of the interfacial membrane lipids, while long-range electrostatic interactions with the NDWCNT enhance the conformational fluctuations of lipid membranes (Li et al. 2012).

One of the most important uses of CNT is as a delivery drug transporter. Using steered molecular dynamics to explore the penetration of CNT encapsulating a polar drug, paclitaxel (PTX), into DPPC bilayer membrane, see Figure 6.15. It is found that Van der Waals interaction between the PTX and the CNT, in addition to the hydrogen bond formation between PTX with the confined water molecules inside CNT, plays an important role in the delivery of the drug (Mousavi et al. 2013). The internalization of short CNT (SWCNT and MWCNT) into a DPPC lipid bilayer was investigated using coarse-grained

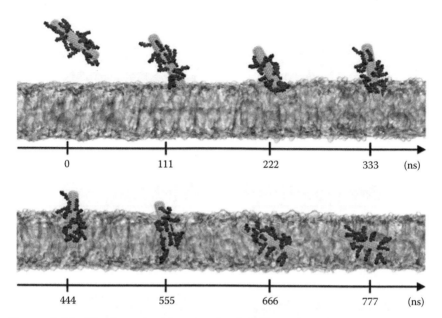

Figure 6.14 Uptake path of functionalized CNT passive diffusion. (Reprinted from *Carbon*, 50(14), Kraszewski, S., Picaud, F., Elhechmi, I., Gharbi, T., and Ramseyer, C., How long a functionalized carbon nanotube can passively penetrate a lipid membrane, 5301–5308, Copyright 2012, with permission from Elsevier.)

and molecular dynamics simulations and it was found that the CNT remain within the lipid bilayer and do not exit spontaneously (Lelimousin and Sansom 2013). One important aspect of this work is that they found the formation of local nonbilayer phases in the lipid bilayer (transbilayer pores and inverted micelles); the pores were transient and the inverted micelles were long lasting. Internalization within lipid membranes by different types of nanotubes include the following: pristine, hydrophobic, homogeneously functionalized, cap functionalized, and terminal rings nanotubes of different lengths and diameters were investigated using molecular dynamics simulations with MARTINI model (Baoukina et al. 2013).

Hydrophobic CNT enter the lipid bilayer spontaneously from water and for different lengths penetration occurs between 100 and 200 ns. The CNT translocate into the membrane and adopt a parallel orientation with respect to the lipid bilayer, see Figure 6.16. Functionalized nanotubes are absorbed on the bilayer surface in the head group region and oriented parallel to the bilayer and this happens in the same period of time as the last example (100–200 ns). CNT with functionalized cap spontaneously enter the bilayer and in this case the CNT remain tilted with their center of mass in the bilayer center. Cap-functionalized and terminal rings nanotubes absorb to the bilayer surface and assume a parallel orientation similar to the functionalized nanotubes but the time required to assume this orientation is much shorter, 6 ns, and the CNT move deeper into the bilayer, see Figure 6.16. One interesting investigation reported in the same work is related with the aggregation of nanotubes and the subsequent translocation within the lipid bilayer. CNT placed separately in water aggregate in 100 ns approximately. Systems containing 4 and 16 CNT of different types were used in this simulation. All aggregates partitioned to the

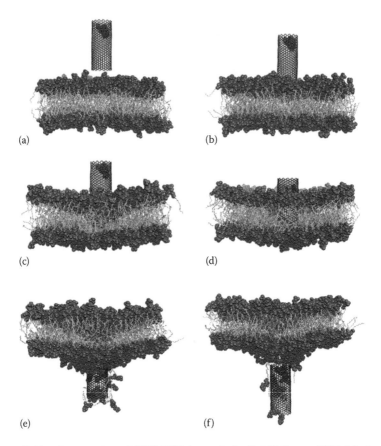

(a)

(b)

(c)

(d)

(e)

(f)

Figure 6.15 Penetration of CNT-PTX through the lipid bilayer. (With kind permission from Springer Science+Business Media: *The Journal of Membrane Biology*, Carbon nanotube-encapsulated drug penetration through the cell membrane: An investigation based on steered molecular dynamics simulation, 246(9), 2013, 697–704, Mousavi, S.Z., Amjad-Iranagh, S., Nademi, Y., and Modarress, H.)

bilayer, either to one or both monolayers and inducing significant perturbation to the system. Lipids from the monolayer closer to the aggregates wrap the hydrophobic sides of the aggregate. This bends the bilayer inducing a strong deformation of the whole bilayer. A later study on CNT bundles insertion within lipid membranes was done using a hybrid particle-field coarse-grained molecular dynamics simulations (Sarukhanyan et al. 2014). They observe a spontaneous insertion of bundles of CNT of different lengths causing distortions in both lipid height and orientation.

Computer simulations of complex molecular events attracted researchers to provide reliable information on cellular processes, such as CNT cellular internalization. In the present time, the interaction of nanostructures with model lipid membranes is possible in simulations. Computer simulation can be used to provide valuable information at molecular level and correct time scale, this information is difficult to access in experimental protocols. In the next section we discuss the main experimental results on the interaction of CNT and lipid membranes.

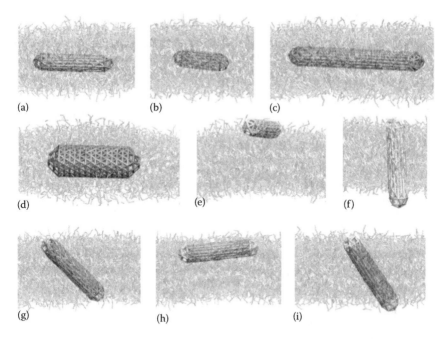

Figure 6.16 NTs inserted into a lipid bilayer: equilibrium and metastable positions and orientations: (a) CNT medium size; (b) CNT short sizes; (c) CNT long size; (d) CNT wide and medium size; functionalized medium size in stable (e) and metastable (f) positions; functionalized caps (g); functionalized caps with terminal rings and medium size in stable (h) and metastable (i) positions. (Reprinted with permission from Baoukina, S., Monticelli, L., and Tieleman, D.P., Interaction of pristine and functionalized carbon nanotubes with lipid membranes, *Journal of Physical Chemistry B*, 117(40), 12113–12123. Copyright 2013 American Chemical Society.)

6.6 LIPID–CNT INTERACTIONS

6.6.1 Self-Assembly

Lipid molecules form half of the mass of most cell membrane (almost all remain mass is protein) and we can estimate that there are about 10^9 lipid molecules in the plasma membrane. Lipids are amphipathic molecules, they have a hydrophilic (polar) section and a hydrophobic (nonpolar) part. Phospholipids are the most abundant lipids in the membrane. They have a polar head group and two hydrophobic tails (usually fatty acids), and these tails can differ in length (from 14 to 24 carbon atoms), where one of the tails is one or more *cis*-double bonds (unsaturated) and the other tail is saturated. Double bonds create a kink in the tail, and this shape and the amphiphilic nature of the lipid molecules cause them to assemble in spherical shape structures called vesicles. In experimental studies, these vesicles can be created at different sizes depending on the methodology used to form them. To create Giant Unilamellar Vesicles (GUVs, with typical sizes ranging from one to thousands of micrometers in diameter), whose sizes are comparable to cells, electroformation or jet-injection methods can be used. In water, the polar head group of the phospholipids forms favorable electrostatic interactions in contrast to the nonpolar tail carbon chains that are completely hydrophobic and hinder from water. On contact with CNT, the nature of the interaction will depend on the type of CNT surface treatment. Next we will discuss the experimental results on CNT–lipid interactions.

The self-assembly of chain lipids with CNT was obtained by simply dissolving the lipids in aqueous solution, followed by CNT incorporation by ultrasonication. This protocol allowed the identification of three possible organization patterns: (1) the perpendicular absorption of the lipids on the CNT surface forming a monolayer; (2) lipids organization in half-cylinders oriented parallel to the tube axis; and (3) half cylinders oriented perpendicular to the tube axis (Richard 2003). Similar results were observed using polar lysophospholipids (Wu et al. 2006). In these experiments, lipid–lipid interactions in the membrane bilayer were obviated resulting in CNT–lipid adsorption in different patterns depending on the CNT inherent properties. This result is important; however, more experimental investigations, as well as computer simulations, are necessary to assess the assembly of lipids on CNT surface within the membrane bilayer.

Amphihilic dendrimers absorbed on the surface of SWCNT in aqueous medium improves the dispersibility of SWCNT. The surface bound dendrimer can be efficiently replaced by small amphiphilies and lysophospholipids (Chang et al. 2010). The stabilization of CNT–lipid assembly by photopolymerization of the diyne motif of the pentose-10,12-diynoic lipids has been obtained (Contal et al. 2010). Reversibility of the self-assembly of photoelectrochemically active lipid bilayer disks arranged parallel to the surface of SWCNT is achieved by subsequent surfactant addition (Boghossian et al. 2011). Self-assembly of glycolipids with SWCNT show specific molecular recognition properties in aqueous solutions (Murthy et al. 2012).

Lipids (SOPS) can be spontaneously wrapped around SWCNT modified with the multilayer polyelectrolyte polymer poly(diallyldimethylammonium chloride) (PDDA), sodium poly(styrenesulphonate) (PSS), and poly(allylamine hydrochloride) (PAH). The layers of polyelecrolyte provide a hydrophilic support surface to maintain the bilayer structure where the lipids are laterally mobile (Artyukhin et al. 2005). COOH-modified SWCNT can be coated with DCPE lipids to form a cylindrical lipid bilayer covering the CNT (He and Urban 2005) and the covalent modification of a COOH-MWCNT can be coated with DPPE lipid and an insertion of α-hemolysin in the lipid-coated MWCNT (Dayani and Malmstadt 2012). CNT can be incorporated in urease-lipid hybrid Langmuir–Blodgett films with a significant modification of the film compounds, creating an appropriate environment to preserve the enzymatic activity of urease, suggesting the use of this system for urea sensing for biomedical applications (Caseli and Siqueira 2012).

6.6.2 Internalization

It has been reported that the interactions of SWCNT and MWCNT with lipids allow then penetrating membranes and cellular plasma membranes through the hydrophobic region of the phospholipids. The proposed mechanism involves the piercing of CNT inducing pore formation (Prylutska et al. 2013). Ion migration across lipid bilayers induced by CNT is monitored in real time with milliseconds temporal sensitivity (Corredor et al. 2013). Aqueous phospholipid membranes (egg-pc and cholesterol) produced at lower salt concentration interact in such a way with CNT that they absorb the lipids instead of piercing the membrane inducing minor perturbations (Shi et al. 2013), and studies cited in that work. Deposition of MWCNT on supported lipid bilayers and vesicles of zwitterionic 1,2-dioleoyl-sn-glycero-3-phosphocholine (DOPC), under neutral pH conditions, increase with increasing electrolyte (NaCl and CaCl$_2$) concentrations; the deposition is irreversible and MWCNT do not disrupt the DOPC bilayers upon attachment (Yi and Chen 2013).

6.7 CONCLUSIONS AND FURTHER PERSPECTIVES

Until now, a great amount of experimental information is available dealing with the effects of CNT on biological systems. This information sometimes is incorrect or very preliminary. There exists controversy between different experimental results and the complexity of the problem makes difficult its elucidation through theoretical or computational models. For instance, in some cases, CNT are considered excellent drug delivery vehicles for a number of drugs, while in other cases CNT are considered toxic for many types of different cells. The identification of the mechanism and the key factors involved in the interaction of CNT with cells and in particular with lipid membranes are paramount for the development of advanced medical applications. There are a number of different unresolved questions. There is an urgent necessity for rigorous research to assess the interaction of carbon nanostructures with cells through the clear formulation of physical and chemical mechanisms. This understanding and the control of the variables influencing CNT–lipid interactions might result in the adequate application of nanotechnology to biological problems without risking human health. The development of innovative, experimental computer simulations and theoretical protocols for the detection, quantification, and qualification of CNT in cells, organs, and organisms is imperative for further use of CNT in biological applications.

ACKNOWLEDGMENTS

The present work was supported by CONACYT and PROMEP through the following project numbers: CB-166014, CB-169504, IDCA 9490-UASLP-CA-214.

REFERENCES

Alarifi, S., D. Ali, A. Verma, F.N. Almajhdi, and A.A. Al-Qahtani. 2014. Single-walled carbon nanotubes induce cytotoxicity and DNA damage via reactive oxygen species in human hepatocarcinoma cells. *In Vitro Cellular & Developmental Biology—Animal* 50 (8): 1–9.

Alloy, M.M. and A.P. Roberts. 2011. Effects of suspended multi-walled carbon nanotubes on daphnid growth and reproduction. *Ecotoxicology and Environmental Safety* 74 (7): 1839–1843.

Artyukhin, A.B., A. Shestakov, J. Harper, O. Bakajin, P. Stroeve, and A. Noy. 2005. Functional one-dimensional lipid bilayers on carbon nanotube templates. *Journal of the American Chemical Society* 127 (20): 7538–7542.

Baoukina, S., L. Monticelli, and D.P. Tieleman. 2013. Interaction of pristine and functionalized carbon nanotubes with lipid membranes. *Journal of Physical Chemistry B* 117 (40): 12113–12123.

Baroli, B. 2010. Penetration of nanoparticles and nanomaterials in the skin: Fiction or reality? *Journal of Pharmaceutical Sciences* 99 (1): 21–50.

Berezhkovskii, A. and G. Hummer. 2002. Single-file transport of water molecules through a carbon nanotube. *Physical Review Letters* 89 (6): 064503.

Bertoncini, P. and O. Chauvet. 2010. Conformational structural changes of bacteriorhodopsin adsorbed onto single-walled carbon nanotubes. *Journal of Physical Chemistry B* 114 (12): 4345–4350.

Bianco, A., K. Kostarelos, C.D. Partidos, and M. Prato. 2005. Biomedical applications of functionalised carbon nanotubes. *Chemical Communications* 5: 571–577.

Boczkowski, J. and S. Lanone. 2012. Respiratory toxicities of nanomaterials—A focus on carbon nanotubes. *Advanced Drug Delivery Reviews* 64 (15): 1694–1699.

Boghossian, A.A., J.H. Choi, M.H. Ham, and M.S. Strano. 2011. Dynamic and reversible self-assembly of photoelectrochemical complexes based on lipid bilayer disks, photosynthetic reaction centers, and single-walled carbon nanotubes. *Langmuir* 27 (5): 1599–1609.

Caseli, L. and J.R. Siqueira. 2012. High enzymatic activity preservation with carbon nanotubes incorporated in urease—Lipid hybrid langmuir—Blodgett films. *Langmuir* 28 (12): 5398–5403.

Chang, D.W., I.-Y. Jeon, J.-B. Baek, and L. Dai. 2010. Efficient dispersion of singlewalled carbon nanotubes by novel amphiphilic dendrimers in water and substitution of the pre-adsorbed dendrimers with conventional surfactants and lipids. *Chemical Communications (Cambridge, England)* 46 (42): 7924–7926.

Chang, R. and A. Violi. 2006. Insights into the effect of combustion-generated carbon nanoparticles on biological membranes: A computer simulation study. *Journal of Physical Chemistry B* 110 (10): 5073–5083.

Contal, E., A. Morère, C. Thauvin, A. Perino, S. Meunier, C. Mioskowski, and A. Wagner. 2010. Photopolymerized lipids self-assembly for the solubilization of carbon nanotubes. *Journal of Physical Chemistry B* 114 (17): 5718–5722.

Contestabile, A. 2001. Oxidative stress in neurodegeneration: Mechanisms and therapeutic perspectives. *Current Topics in Medicinal Chemistry (Netherlands)* 1 (6): 553–568.

Corredor, C., W.-C. Hou, S.A. Klein, B.Y. Moghadam, M. Goryll, K. Doudrick, P. Westerhoff, and J.D. Posner. 2013. Disruption of model cell membranes by carbon nanotubes. *Carbon* 60: 67–75.

Dayani, Y. and N. Malmstadt. 2012. Lipid bilayers covalently anchored to carbon nanotubes. *Langmuir* 28 (21): 8174–8182.

Fischer, H.C. and W.C.W. Chan. 2007. Nanotoxicity: The growing need for in vivo study. *Current Opinion in Biotechnology* 18 (6): 565–571.

Garate, J.A., N.J. English, and J.M.D. MacElroy. 2009a. Carbon nanotube assisted water self-diffusion across lipid membranes in the absence and presence of electric fields. *Molecular Simulation* 35 (November 2013): 3–12.

Garate, J.-A., N.J. English, and J.M.D. MacElroy. 2009b. Static and alternating electric field and distance-dependent effects on carbon nanotube-assisted water self-diffusion across lipid membranes. *Journal of Chemical Physics* 131 (11): 114508.

Gasser, M., B. Rothen-Rutishauser, H.F. Krug, P. Gehr, M. Nelle, B. Yan, and P. Wick. 2010. The adsorption of biomolecules to multi-walled carbon nanotubes is influenced by both pulmonary surfactant lipids and surface chemistry. *Journal of Nanobiotechnology* 8 (1): 31.

Gasser, M., P. Wick, M.J.D. Clift, F. Blank, L. Diener, B. Yan, P. Gehr, H.F. Krug, and B. Rothen-Rutishauser. 2012. Pulmonary surfactant coating of multi-walled carbon nanotubes (MWCNTs) influences their oxidative and pro-inflammatory potential in vitro. *Particle and Fibre Toxicology* 9 (1): 17.

He, P. and M.W. Urban. 2005. Controlled phospholipid functionalization of single-walled carbon nanotubes. *Biomacromolecules* 6: 2455–2457.

Höfinger, S., M. Melle-Franco, T. Gallo, A. Cantelli, M. Calvaresi, J.A.N.F. Gomes, and F. Zerbetto. 2011. A computational analysis of the insertion of carbon nanotubes into cellular membranes. *Biomaterials* 32 (29): 7079–7085.

Hummer, G., J.C. Rasaiah, and J.P. Noworyta. 2001. Water conduction through the hydrophobic channel of a carbon nanotube. *Nature* 414 (6860): 188–190.

Jain, S., V.S. Thakare, M. Das, C. Godugu, A.K. Jain, R. Mathur, K. Chuttani, and A.K. Mishra. 2011. Toxicity of multiwalled carbon nanotubes with end defects critically depends on their functionalization density. *Chemical Research in Toxicology* 24 (11): 2028–2039.

Jos, A., S. Pichardo, M. Puerto, E. Sánchez, A. Grilo, and A.M. Cameán. 2009. Cytotoxicity of carboxylic acid functionalized single wall carbon nanotubes on the human intestinal cell line Caco-2. *Toxicology in Vitro* 23 (8): 1491–1496.

Kagan, V.E., Y.Y. Tyurina, V.A. Tyurin, N.V. Konduru, A.I. Potapovich, A.N. Osipov, E.R. Kisin et al. 2006. Direct and indirect effects of single walled carbon nanotubes on RAW 264.7 macrophages: Role of iron. *Toxicology Letters* 165 (1): 88–100.

Kapralov, A.A., W.H. Feng, A.A. Amoscato, N. Yanamala, K. Balasubramanian, D.E. Winnica, E.R. Kisin et al. 2012. Adsorption of surfactant lipids by single-walled carbon nanotubes in mouse lung upon pharyngeal aspiration. *ACS Nano* 6 (5): 4147–4156.

Kraszewski, S., F. Picaud, I. Elhechmi, T. Gharbi, and C. Ramseyer. 2012. How long a functionalized carbon nanotube can passively penetrate a lipid membrane. *Carbon* 50 (14): 5301–5308.

Lacerda, L., S. Raffa, M. Prato, A. Bianco, and K. Kostarelos. 2007. Cell-penetrating CNTs for delivery of therapeutics. *Nano Today* 2 (6): 38–43.

Lelimousin, M. and M.S.P. Sansom. 2013. Membrane perturbation by carbon nanotube insertion: Pathways to internalization. *Small* 9: 3639–3646.

Li, Z., T. Hulderman, R. Salmen, R. Chapman, S.S. Leonard, S.H. Young, A. Shvedova, M.I. Luster, and P.P. Simeonova. 2007. Cardiovascular effects of pulmonary exposure to single-wall carbon nanotubes. *Environmental Health Perspectives* 115 (3): 377–382.

Li, X., Y. Shi, B. Miao, and Y. Zhao. 2012. Effects of embedded carbon nanotube on properties of biomembrane. *Journal of Physical Chemistry B* 116 (18): 5391–5397.

Lim, J.-H., S.-H. Kim, I.-S. Shin, N.-H. Park, C. Moon, S.-S. Kang, S.-H. Kim, S.-C. Park, and J.-C. Kim. 2011. Maternal exposure to multi-wall carbon nanotubes does not induce embryo-fetal developmental toxicity in rats. *Birth Defects Research. Part B, Developmental and Reproductive Toxicology* 92 (1): 69–76.

Micoli, A., M.L. Soriano, H. Traboulsi, M. Quintana, and M. Prato. 2013. ZnII-cyclen as a supramolecular probe for tagging thymidine nucleosides on carbon nanotubes. *European Journal of Organic Chemistry* 18: 3685–3690.

Micoli, A., A. Turco, E. Araujo-Palomo, A. Encinas, M. Quintana, and M. Prato. 2014. Supramolecular assembles of nucleoside functionalized carbon nanotubes: Synthesis, film preparation, and properties. *Chemistry A European Journal* 20: 5397–5402.

Monticelli, L., E. Salonen, P.C. Ke, and I. Vattulainen. 2009. Effects of carbon nanoparticles on lipid membranes: A molecular simulation perspective. *Soft Matter* 5 (22): 4433.

Mousavi, S.Z., S. Amjad-Iranagh, Y. Nademi, and H. Modarress. 2013. Carbon nanotube-encapsulated drug penetration through the cell membrane: An investigation based on steered molecular dynamics simulation. *The Journal of Membrane Biology* 246 (9): 697–704.

Murthy, B.N., S. Zeile, M. Nambiar, M.R. Nussio, C.T. Gibson, J.G. Shapter, N. Jayaraman, and N.H. Voelcker. 2012. Self assembly of bivalent glycolipids on single walled carbon nanotubes and their specific molecular recognition properties. *RSC Advances* 2: 1329.

Nel, A.E., L. Madler, D. Velegol, T. Xia, E.M.V. Hoek, P. Somasundaran, F. Klaessig, V. Castranova, and M. Thompson. 2009. Understanding biophysicochemical interactions at the nano-bio interface. *Nature Materials* 8 (7): 543–557.

Oberdörster, G., E. Oberdörster, and J. Oberdörster. 2005. Nanotoxicology: An emerging discipline evolving from studies of ultrafine particles. *Environmental Health Perspectives* 113 (7): 823–839.

Park, J., D. Hong, D. Kim, K. Byun, and S. Hong. 2014. Anisotropic membrane diffusion of human mesenchymal stem cells on aligned single-walled carbon nanotube networks. *The Journal of Physical Chemistry C* 118 (7): 3742–3749.

Parthasarathi, R., N.R. Tummala, and A. Striolo. 2012. Embedded single-walled carbon nanotubes locally perturb DOPC phospholipid bilayers. *Journal of Physical Chemistry B* 116 (42): 12769–12782.

Patra, N. and P. Král. 2011. Controlled self-assembly of filled micelles on nanotubes. *Journal of the American Chemical Society* 133 (16): 6146–6149.

Pogodin, S. and V.A. Baulin. 2010. Can a carbon nanotube pierce through a phospholipid bilayer? *ACS Nano* 4 (9): 5293–5300.

Prato, M., K. Kostarelos, and A. Bianco. 2008. Functionalized carbon nanotubes in drug design and discovery. *Accounts of Chemical Research* 41 (1): 60–68.

Prylutska, S., R. Bilyy, T. Shkandina, D. Rotko, A. Bychko, V. Cherepanov, R. Stoika et al. 2013. Comparative study of membranotropic action of single- and multi-walled carbon nanotubes. *Journal of Bioscience and Bioengineering* 115 (6): 674–679.

Raposo, G. and W. Stoorvogel. 2013. Extracellular vesicles: Exosomes, microvesicles, and friends. *Journal of Cell Biology* 200 (4): 373–383.

Ravichandran, P., S. Baluchamy, B. Sadanandan, R. Gopikrishnan, S. Biradar, V. Ramesh, J.C. Hall, and G.T. Ramesh. 2010. Multiwalled carbon nanotubes activate NF-κB and AP-1 signaling pathways to induce apoptosis in rat lung epithelial cells. *Apoptosis* 15 (12): 1507–1516.

Ravichandran, P., A. Periyakaruppan, B. Sadanandan, V. Ramesh, J.C. Hall, O. Jejelowo, and G.T. Ramesh. 2009. Induction of apoptosis in rat lung epithelial cells by multiwalled carbon nanotubes. *Journal of Biochemical and Molecular Toxicology* 23 (5): 333–344.

Reddy, A.R.N., Y.N. Reddy, D.R. Krishna, and V. Himabindu. 2010. Multi wall carbon nanotubes induce oxidative stress and cytotoxicity in human embryonic kidney (HEK293) cells. *Toxicology* 272 (1–3): 11–16.

Reddy, A.R.N., M. Venkateswar Rao, D.R. Krishna, V. Himabindu, and Y. Narsimha Reddy. 2011. Evaluation of oxidative stress and anti-oxidant status in rat serum following exposure of carbon nanotubes. *Regulatory Toxicology and Pharmacology: RTP* 59 (2): 251–257.

Richard, C. 2003. Supramolecular self-assembly of lipid derivatives on carbon nanotubes. *Science* 300 (5620): 775–778.

Riding, M.J., F.L. Martin, J. Trevisan, V. Llabjani, I.I. Patel, K.C. Jones, and K.T. Semple. 2012. Concentration-dependent effects of carbon nanoparticles in gram-negative bacteria determined by infrared spectroscopy with multivariate analysis. *Environmental Pollution* 163: 226–234.

Roberts, A.P., A.S. Mount, B. Seda, J. Souther, R. Qiao, S. Lin, C.K. Pu, A.M. Rao, and S.J. Klaine. 2007. In vivo biomodification of lipid-coated carbon nanotubes by *Daphnia magna Environmental Science and Technology* . 41 (8): 3028–3029.

Sarukhanyan, E., A. De Nicola, D. Roccatano, T. Kawakatsu, and G. Milano. 2014. Spontaneous insertion of carbon nanotube bundles inside biomembranes: A hybrid particle-field coarse-grained molecular dynamics study. *Chemical Physics Letters* 595-596: 156–166.

Shao, W., A. Paul, B. Zhao, C. Lee, L. Rodes, and S. Prakash. 2013. Carbon nanotube lipid drug approach for targeted delivery of a chemotherapy drug in a human breast cancer xenograft animal model. *Biomaterials* 34 (38): 10109–10119.

Shi, L., D. Shi, M.U. Nollert, D.E. Resasco, and A. Striolo. 2013. Single-walled carbon nanotubes do not pierce aqueous phospholipid bilayers at low salt concentration. *Journal of Physical Chemistry B* 117 (22): 6749–6758.

Shvedova, A.A., E.R. Kisin, A.R. Murray, O. Gorelik, S. Arepalli, V. Castranova, S.H. Young et al. 2007. Vitamin E deficiency enhances pulmonary inflammatory response and oxidative stress induced by single-walled carbon nanotubes in C57BL/6 mice. *Toxicology and Applied Pharmacology* 221 (3): 339–348.

Shvedova, A.A., A. Pietroiusti, B. Fadeel, and V.E. Kagan. 2012. Mechanisms of carbon nanotube-induced toxicity: Focus on oxidative stress. *Toxicology and Applied Pharmacology* 261 (2): 121–133.

Skandani, A.A., R. Zeineldin, and M. Al-Haik. 2012. Effect of chirality and length on the penetrability of single-walled carbon nanotubes into lipid bilayer cell membranes. *Langmuir: The ACS Journal of Surfaces and Colloids* 28 (20): 7872–7879.

Tsukahara, T. and H. Haniu. 2011. Nanoparticle-mediated intracellular lipid accumulation during C2C12 cell differentiation. *Biochemical and Biophysical Research Communications* 406 (4): 558–563.

Tyurina, Y.Y., E.R. Kisin, A. Murray, V.A. Tyurin, V.I. Kapralova, L.J. Sparvero, A.A. Amoscato et al. 2011. Global phospholipidomics analysis reveals selective pulmonary peroxidation profiles upon inhalation of single-walled carbon nanotubes. *ACS Nano* 5 (9): 7342–7353.

Wallace, E.J. and M.S.P. Sansom. 2009. Carbon nanotube self-assembly with lipids and detergent: A molecular dynamics study. *Nanotechnology* 20 (4): 045101.

Wang, H., S. Michielssens, S.L.C. Moors, and A. Ceulemans. 2009. Molecular dynamics study of dipalmitoylphosphatidylcholine lipid layer self-assembly onto a single-walled carbon nanotube. *Nano Research* 2 (12): 945–954.

Wang, J., P. Sun, Y. Bao, B. Dou, D. Song, and Y. Li. 2012. Vitamin E renders protection to PC12 cells against oxidative damage and apoptosis induced by single-walled carbon nanotubes. *Toxicology in Vitro: An International Journal Published in Association with BIBRA* 26 (1): 32–41.

Wu, Y., J.S. Hudson, Q. Lu, J.M. Moore, A.S. Mount, A.M. Rao, E. Alexov, and P.C. Ke. 2006. Coating single-walled carbon nanotubes with phospholipids. *Journal of Physical Chemistry B* 110 (6): 2475–2478.

Yi, P. and K.L. Chen. 2013. Interaction of multiwalled carbon nanotubes with supported lipid bilayers and vesicles as model biological membranes. *Environmental Science & Technology* 47 (11): 5711–5719.

Zhang, Y., Y. Xu, Z. Li, T. Chen, S.M. Lantz, P.C. Howard, M.G. Paule et al. 2011. Mechanistic toxicity evaluation of uncoated and PEGylated single-walled carbon nanotubes in neuronal PC12 cells. *ACS Nano* 5 (9): 7020–7033.

Zhornik, E.V., L.A. Baranova, A.M. Strukova, E.N. Loiko, and I.D. Volotovski. 2012. ROS induction and structural modification in human lymphocyte membrane under the influence of carbon nanotubes. *Biophysics* 57 (3): 325–331.

Zhu, B., X. Xia, N. Xia, S. Zhang, and X. Guo. 2014. Modification of fatty acids in membranes of bacteria: Implication for an adaptive mechanism to the toxicity of carbon nanotubes. *Environmental Science and Technology* 48 (7): 4086–4095.

PART III
MODELING IN BIOMATERIALS

7 Computational Model–Driven Design of Tissue-Engineered Vascular Grafts

Ramak Khosravi, Christopher K. Breuer,
Jay D. Humphrey, and Kristin S. Miller

CONTENTS

7.1 CLINICAL NEED, CHALLENGES, AND MOVING TOWARD RATIONAL DESIGN

Over 600,000 patients per year, require vascular grafts for bypass procedures, hemodialysis, or repair of congenital heart defects in the United States alone (AHA 2015). Unfortunately, such interventions are limited by either the lack of availability of autologous vessels or the suboptimal performance of widely available synthetic vascular grafts (Kannan et al. 2005). In particular, current synthetic grafts lack growth potential and they continue to be prone to complications such as thrombosis, stenosis, dilatation, and infection, all of which can necessitate revision procedures. These complications result primarily from the long-term presence of nonliving biomaterials such as Dacron and ePTFE (Seifu et al. 2013).

To overcome the limitations of synthetic grafts, many research groups have turned to tissue engineering to develop conduits made from autologous cells (Hibino et al. 2010, Dahl et al. 2011, Wystrychowski et al. 2014). A now longstanding approach is to tissue engineer a living vessel *ex vivo* by seeding either biodegradable scaffolds or decellularized native tissue with autologous cells and then incubating these constructs within sophisticated bioreactors until a neovessel forms (see, in particular, the seminal publications by Niklason et al. (1999) and L'Heureux et al. (2006)). Although such approaches have advanced to clinical trials, they can be limited by long lead times and high production costs (Seifu et al. 2013). In an attempt to reduce costs and enable off-the-shelf availability, other groups have directly implanted either cell-free or cell-seeded biodegradable polymeric scaffolds within patients (Hibino et al. 2010). The obvious advantage of this approach is that the scaffolds can have off-the-shelf availability and thus be ready for implantation on short notice. The challenge, however, is to ensure that the construct retains structural integrity throughout the *in vivo* process of scaffold degradation and neotissue formation. This approach has similarly advanced to clinical trials, particularly as extracardiac cavopulmonary conduits in children with single ventricle physiology (cf. Hibino et al. 2010), and will be the focus of this chapter.

Most advances in the field of tissue engineering over the past 20 years, regardless of specific approach or application, have been realized by trial-and-error, mainly from the painstaking empirical evaluation of myriad construct biomaterials, cell sources, culture conditions, and so forth. Moreover, most prior preclinical studies were necessarily concerned with basic safety and efficacy, not long-term functionality and performance. For example, the initial focus on tissue-engineered vascular grafts (TEVGs) was to avoid acute catastrophic graft failure due to dehiscence or graft rupture. Subsequently, focus turned to minimizing short-term dilatation, acute thrombosis, or rupture. Notwithstanding the continued importance of these basic issues, recent successes with scaffolds that have appropriate initial mechanical properties for implantation in the low-pressure venous circulation have encouraged a paradigm shift toward the optimization of long-term mechanobiological stability and growth potential (Roh et al. 2008, Hibino et al. 2011, Naito et al. 2013, Khosravi et al. 2015). Achieving this far more ambitious outcome will require, among other things, identification of optimal scaffold structure and material properties to yield the desired *in vivo* functionality (Miller et al. 2015). Hence, there has been a call for a more rational design of TEVGs (Hibino et al. 2011), and computational biomechanical models promise to aid in this pursuit (Niklason et al. 2010, Miller et al. 2014).

The first computational biomechanical model of TEVG development captured collagenous tissue growth within poly(glycol acid) (PGA) scaffolds seeded with smooth muscle cells in a bioreactor (Niklason et al. 2010). This model accounted both for the kinetics of polymer degradation and the synthesis and degradation of multiple families of collagen fibers in response to cyclic strains imparted by luminal pressure. Simulations predicted well both the evolving overall thickness and stress–stretch relationships for tubular engineered vessels cultured for 8 weeks, and concurrent nonlinear optical microscopy revealed that collagen fiber alignment was driven strongly by nondegraded polymer fibers at early times during culture, with subsequent mechano-stimulated dispersal of fiber orientations as polymer fibers degraded. This study showed for the first time that computational models of the growth and remodeling of polymer-based engineered vessels can capture salient aspects of collagen matrix development *in vitro* and can predict evolving tissue morphology and mechanics over long culture periods.

7.2 MURINE MODELS FOR SMALL-CALIBER VASCULAR GRAFTS

To mimic the low-pressure mechanical environment of the extracardiac total cavopulmonary connection in children following the Fontan procedure (Patterson et al. 2012), and to accelerate preclinical evaluation and screening of small-diameter vascular conduits *in vivo*, Roh et al. (2008) developed a murine model wherein polymeric scaffolds are implanted in the venous circulation. Tubular scaffolds were constructed from sheets of nonwoven PGA, sealed with a 50:50 copolymer solution composed of poly(e-caprolactone and l-lactide) (P(CL/LA)), and implanted as interposition vascular grafts in the mouse inferior vena cava (IVC) (Figure 7.1a). This murine model recapitulates many aspects of vascular neotissue formation described in both pediatric patients and previous large animal work, although changes occur over a shorter period (Roh et al. 2010). It also allowed a detailed longitudinal study of cellular and molecular mechanisms underlying neotissue formation, including the important discovery that transformation *in vivo* of a scaffold into a living vascular conduit is driven primarily by an immune-mediated process orchestrated by infiltrating host monocytes (Figure 7.1c) (Roh et al. 2008, 2010, Hibino et al. 2011b). These inflammatory cells, in turn, recruit smooth muscle cells from the neighboring vascular wall and allow luminal endothelialization via a paracrine mechanism (Hibino et al. 2011a). The recruited mesenchymal cells appropriately synthesize and organize a prototypical vascular structure while the monocytes progressively diminish as the scaffold degrades; in this way, these cooperating cells create a completely autologous neovessel. Longitudinal studies revealed that these TEVGs remained patent for 24 weeks, with no evidence of thrombosis, stenosis, or dilatation (Naito et al. 2013). A custom biaxial mechanical testing device (Gleason et al. 2004)

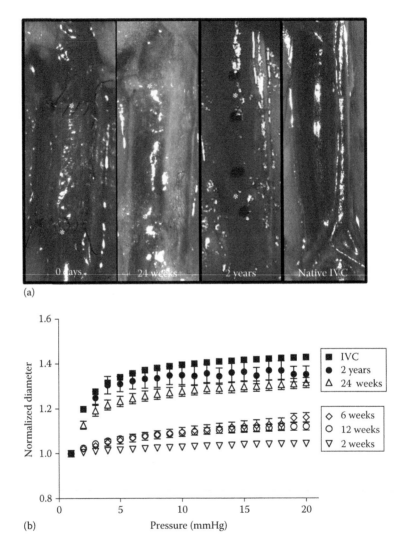

(a)

(b)

Figure 7.1 (a) Gross images of TEVGs implanted as interposition grafts in the murine IVC: 0 days, 24 weeks, and 2 years after implantation with the native IVC shown for comparison. (b) *In vitro* pressure-diameter responses show a gradual increase in distensibility of the TEVG throughout the implantation period. There is a transition from a stiff scaffold-dominated mechanical behavior at 2 weeks to a more distensible, neotissue-dominated mechanical behavior that resembles the native IVC at 2 years. (*Continued*)

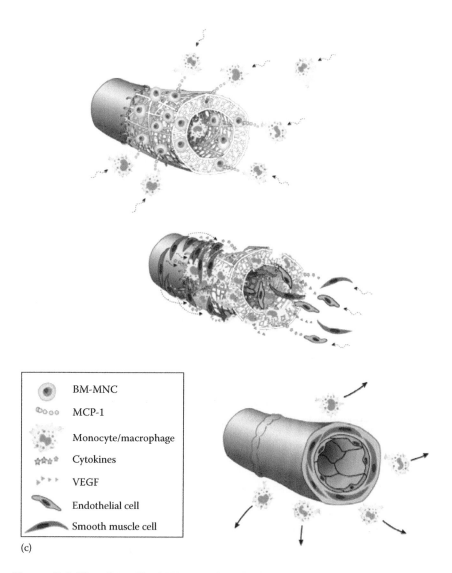

BM-MNC	
MCP-1	
Monocyte/macrophage	
Cytokines	
VEGF	
Endothelial cell	
Smooth muscle cell	

(c)

Figure 7.1 (*Continued*)   (c) Proposed mechanism of vascular transformation of seeded biodegradable scaffolds into autologous neovessels. Infiltrating monocytes release multiple chemokines, cytokines, and growth factors (most notably, monocyte chemoattractant protein 1 [MCP-1]), which recruit smooth muscle cells and endothelial cells that appropriately organize into a blood vessel structure. (Modified from Khosravi, R. et al., *Tissue Eng. A*, 21, 1529, 2015; Reprinted with permission of Roh, J.D. et al., *Proc. Natl. Acad. Sci.*, 107, 4669, 2010.)

was used to quantify *in vitro* the circumferential and axial mechanical properties of grafts explanted at multiple times. Given that PGA is largely degraded by week 6, and nearly absent at week 12, these tests confirmed that the mechanical properties of the neovessel evolve such that they are characterized less-and-less by polymer-dominated stiffness, with a progressive increase in distensibility that approaches that of the native IVC at week 24 (Figure 7.1b) (Naito et al. 2013). Neovessel development reached a steady remodeled state by 24 weeks that remained nearly unchanged for 2 years—essentially, the lifespan of the mouse (Khosravi et al. 2015).

In an effort to transition this concept to the arterial circulation, Udelsman et al. (2014) initially turned to polymer tubular scaffolds fabricated from poly(lactic acid) (PLA) and coated with a 50:50 copolymer of P(CL/LA) (Roh et al. 2008). Following seeding with syngeneic bone marrow–derived mononuclear cells, these scaffolds were implanted as infrarenal aortic interposition grafts in wild-type mice and monitored serially using ultrasound (see Figure 7.2 for a comparison of these arterial grafts with the venous design). Similar to the studies on the venous TEVGs, grafts were explanted after long periods of *in vivo* development (e.g., 3 or 7 months). In this case, however, the TEVGs remained much stiffer biaxially than the native tissue, suggesting that the high-stiffness PLA has inadequate *in vivo* degradation, which impairs cell-mediated development of vascular neotissue having properties closer to native arteries (Udelsman et al. 2014). To reduce chronic inflammation due to the long-term persistence of the polymer, Khosravi et al. (2016) developed a novel method for electrospinning TEVGs composed of a poly(glycerol sebacate) (PGS) microfibrous core enveloped by a thin poly(caprolactone) (PCL) sheath and implanted them as infrarenal aortic interposition grafts in mice for up to 1 year, with favorable outcomes. The thicker, rapidly degrading inner layer of PGS models a distensible "media" that promotes infiltration of host inflammatory and synthetic cells, and the thinner, long-lasting outer PCL nanofiber sheath models a protective "adventitia" that provides mechanical support. This work was built on the past success of cell-free elastomeric grafts containing solvent-casted, particulate-leached (SCPL) PGS that degraded rapidly and promoted neoartery development in a rat model (Wu et al. 2012). Compared with SCPL fabrication, electrospinning improves suture retention, ultimate tensile strength, and strain to failure, while also improving fabrication throughput and reproducibility (Jeffries et al. 2015). Findings suggest that although occupying a small percentage of the original cross-sectional area, persistence of the PCL, up to 1 year, limits full neovessel development. There is a need, therefore, to optimize both the core and the sheath to promote efficient neovessel development.

7.3 COMPUTATIONAL MODEL FOR *IN VIVO* NEOVESSEL DEVELOPMENT

Motivated by data on the evolution of microstructural composition and the mechanical properties of the aforementioned mouse IVC interposition grafts over 24 weeks (Naito et al. 2013), we developed a first-generation theoretical framework for describing the *in vivo* development of a tissue-engineered vein from our implanted polymeric scaffold (Miller et al. 2014). Our growth and remodeling (G&R) approach allows a classical formulation of the wall mechanics, including equilibrium and constitutive relations for stress: div $\mathbf{t} = \mathbf{0}$ with Cauchy stress $(\det \mathbf{F})\mathbf{t} = 2\mathbf{F} \dfrac{\partial W}{\partial \mathbf{C}} \mathbf{F}^T$, where $\mathbf{C} = \mathbf{F}^T\mathbf{F}$ is the right Cauchy–Green tensor (Humphrey 2002, Humphrey and Rajagopal 2002). This approach admits a simple "rule of mixtures" formulation; hence, the energy stored, W, in the TEVG is assumed to equal the sum of the energies stored in all structurally significant constituents, W^α (with $\alpha = p$ (polymer), $1, \ldots, n$ (matrix)), each of

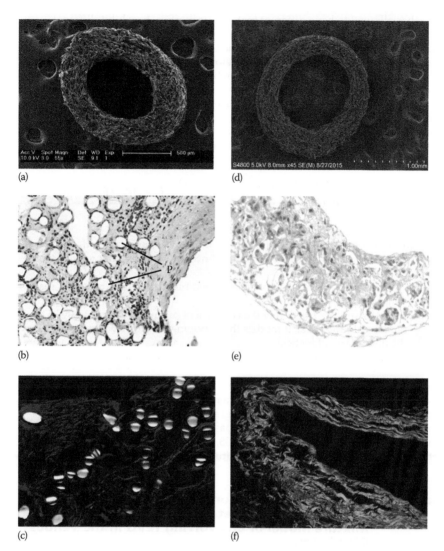

Figure 7.2 Scanning electron microscopy (a and d), H&E staining (b and e), and Picrosirius red staining (c and f) of PLA grafts (left) and PGA-P(CL/LA) grafts (right) implanted as abdominal aortic and venous IVC interposition grafts in the murine circulation, respectively. Histological images are taken at 40× magnification. Note the abundant residual polymer (labeled "P") in the PLA graft at 3 months, which is absent in the PGA-P(CL/LA) graft comprised mostly of neotissue at the same time. Also note the significantly greater quantity of thick, densely packed collagen fibers (red) encapsulating polymer remnants in the arterial graft at 7 months, in contrast to the noticeably more thin, loosely packed collagen (green and yellow) comprising the neotissue in the venous graft at 2 years. (Modified from Udelsman, B.V., *J. Biomech.*, 47, 2070, 2014; SEM images: Courtesy of Kevin Rocco (Yale University, New Haven, CT) and Cameron Best (Nationwide Children's Hospital, Columbus, OH), histological images courtesy of Brooks Udelsman and Yuji Naito (Yale University, New Haven, CT).)

which is allowed to have different material properties, to exhibit different rates of production and removal, and to possess different natural (stress-free) configurations, yet they are all constrained to deform with the bulk material (Humphrey and Rajagopal 2002). The deformations experienced by individual constituents can thus be quantified given both the gross deformations of the composite TEVG, which are measurable *in vivo*, and deformations associated with the incorporation of individual matrix constituents within the extant material. Together, these two deformations yield the constituent-specific deformation gradients $\mathbf{F}_{n(\tau)}^{\alpha}(s)$ at any G&R time s, where $n(\tau)$ denotes the constituent-specific natural configuration and $\tau \in [0, s]$ the time of deposition during the G&R simulation. Therefore, $W(s) = \sum W^{\alpha}(s)$, where

$$W^{\alpha}(s) = \frac{\rho^{P}(0)Q^{P}(s)}{\rho^{P}(s)}\hat{W}^{P}\left(\mathbf{F}_{n(0)}^{P}(s)\right) + \int_{0}^{s}\frac{m^{\alpha}(\tau)}{\rho(s)}q^{\alpha}(s-\tau)\hat{W}^{\alpha}\left(\mathbf{F}_{n(\tau)}^{\alpha}(s)\right)d\tau,$$

where
- $\rho^{P}(0)$ is the initial apparent mass density of the polymeric scaffold
- $Q^{P}(s) \in [0, 1]$ is the fraction of the scaffold present at time 0 that remains at the current time s
- $m^{\alpha}(\tau)$ is the extracellular matrix (ECM) constituent-specific rate of new mass density production
- $q^{\alpha}(s-\tau) \in [0, 1]$ is the fraction of the constituent produced at time $\tau \in [0, s]$ that remains at time s (i.e., it models the surviving fraction by accounting for the degradation kinetics)
- $W^{\alpha}\left(\mathbf{F}_{n(\tau)}^{\alpha}(s)\right)$ is the stored energy function of each constituent

Mechanical contributions of the polymeric scaffold and constituents of the extracellular matrix are described using a neo-Hookean relation for the scaffold and amorphous extracellular matrix and Fung-type exponentials for the multiple families of collagen fibers and passive smooth muscle, all motivated by data for the native vein (Roh et al. 2008, Sokolis 2012, Lee et al. 2013, Naito et al. 2013) (Table 7.1). The specific constitutive relations used to describe and

Table 7.1: Specific Functional Forms of Constitutive Relations Employed by Miller et al. (2014) in a Constrained Mixture Model of an Evolving TEVG in the Venous Circulation

Class	Relationship	Form
Well accepted	Mass density removal–time (neotissue)	Exponential decay
	Energy density–strain (collagen/muscle)	Fung exponential
	Energy density–strain (polymer)	Neo-Hookean
Postulated	Mass density production–stress	Linear
	Mass density production–inflammation	Gamma
	Damage function–polymer	Linear after P(CL/LA) coating degrades
	Mass density removal–time (polymer)	Sigmoidal

Source: Miller, K.S. et al., *J. Biomech.*, 47, 2080, 2014.
Note: Well-accepted relations are those supported by experimental data (Valentin and Humphrey 2009, Niklason et al. 2010), while postulated forms represent those for which such data were absent.

predict the evolution of TEVGs in the murine venous circulation capture the degradation and loss of load-bearing integrity of the initial polymeric scaffold and the subsequent turnover of the extracellular matrix, which is driven by an early foreign body inflammatory response and a subsequent mechano-mediated response coordinated by mechanosensitive synthetic cells (Hibino et al. 2011a,b). Briefly, matrix production is modeled as,

$$m^k(\tau) = m_B^k\left(1 - e^{-\tau}\right)\left(m_{infl}^k(\tau) + m_{strs}^k(\tau) + 1\right),$$

where
m_B^k is the basal rate of production in the native IVC at a homeostatic state of stress

$(1 - e^{-\tau})$ accounts for the delayed cellular infiltration into the scaffold and the associated initiation of matrix production

m_{infl}^k and m_{strs}^k represent inflammation- and stress-mediated production, respectively

Specifically, $m_{infl}^k(\tau) \equiv K_{infl}(\alpha\tau e^{1-\alpha\tau})$, where K_{infl} is a rate parameter representing increases in synthetic capability and $1/\alpha$ represents the peak time of monocyte infiltration (Hibino et al. 2011b); $m_{strs}^k(\tau) \equiv \left(1 - \left(c^p(\tau)/c^p(0)\right)\right)K_{strs}^k\Delta\sigma^k(\tau)$, where $(1 - (c^p(\tau)/c^p(0))) \in [0,1]$ captures the initial stress-shielding of mechanosensitive cells by a stiff polymer, K_{strs}^k is a rate-type parameter, and $\Delta\sigma^k(\tau)$ is a normalized difference in intramural stress from a preferred target ("homeostatic") value (Valentin and Humphrey 2009). To model the loss of matrix due to basal time-, tension-, and inflammation-dependent degradation, we let fibrillar collagen degrade via a first-order kinetic decay:

$$q^k(s,\tau) = \exp\left(-\int_\tau^s k^k(\hat{\tau})d\hat{\tau}\right),$$

where $k^k(\hat{\tau})$ is a rate parameter having units of days^{-1}. This general G&R framework was used to simulate TEVGs grown *in vitro* in bioreactors (Niklason et al. 2010), but the addition of the inflammation-mediated component captures well the *in vivo* evolving mechanical behaviors of TEVGs implanted as interposition grafts in the murine IVC (Figure 7.3, Miller et al. 2014). Given sufficient experimental data, we submit that this general framework can similarly be modified and extended to predict the long-term performance of bilayered vascular grafts in the arterial circulation (Khosravi et al. 2016).

7.3.1 Considerations When Modeling the Polymer

To ensure that their PGA-P(CL/LA) scaffolds possessed sufficient mechanical strength, Roh et al. (2008) evaluated the initial suture retention strength, burst pressure, and elastic modulus, then followed the tensile strength of the scaffolds over a 6-month period *in vitro* to assess loss of mechanical strength due to hydrolytic degradation. Although the scaffolds demonstrated appropriate initial mechanical properties that either exceeded the physiologic range of native vessels or outperformed synthetic graft materials such as ePTFE, the authors appropriately noted that these properties deteriorated when exposed to the hemodynamic and inflammatory environments of the circulatory system, which in turn influenced the stresses and strains experienced by infiltrating cells. That is, *in vitro* tests are necessary prerequisites, but they must be followed by long-term *in vivo* studies. Indeed,

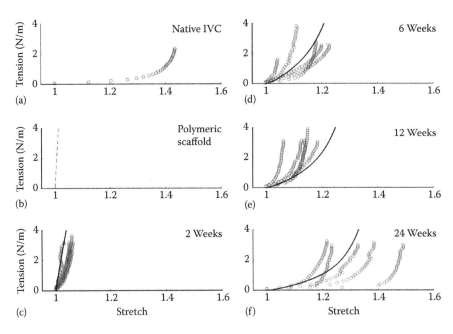

Figure 7.3 Circumferential tension–stretch relationships for the native murine IVC (a), simulated polymeric scaffold at time 1 day (b), and both predicted (solid curves) and experimental (circles) results for TEVGS at 2 (c), 6 (d), 12 (e), and 24 (f) weeks. The mechanical behavior (predicted and experimental) evolved from a stiff response dominated by polymer and inflammation-mediated collagen (c) to a compliant behavior due primarily to collagen produced during the stress-mediated phase (f). Despite the variation in experimental data, model predictions agreed well observed with the mechanical behavior observed during biaxial testing. (Modified from Miller, K.S. et al., *J. Biomech.*, 47, 2080, 2014; Khosravi, R. et al., *Tissue Eng. A*, 21, 1529, 2015.)

despite some limitations, Roh et al. (2008) confirmed that their grafts could, at least over short periods, withstand pressurization and *in vivo* loading conditions in the venous circulation.

Motivated by *in vitro* findings by Niklason et al. (2010) and *in vivo* findings by Roh et al. (2008) and Naito et al. (2013), we sought to develop an associated computational, that is, *in silico*, model (Miller et al. 2014). It was assumed that the stored energy function associated with the PGA-P(CL/LA) polymeric scaffold was of a neo-Hookean form (Niklason et al. 2010), with the initial value of the shear modulus identified from uniaxial tensile tests performed in a pilot study. Recent studies have similarly reported biaxial mechanical properties and corresponding constitutive relations for polymers of interest for their respective vascular applications (Hu 2015, Tamini et al. 2016). Biaxial tests are preferred, of course, for the *in vivo* loading is biaxial; thus, there is a need to model scaffold stiffness based on such data.

In addition, it is important to account for the kinetics of polymer degradation. When one is modeling the loss of a polymer, a distinction must be made between the loss of mass and the loss of structural integrity. Although residual polymer fragments may persist for some time in the graft wall and elicit a chronic inflammatory response, their load-bearing integrity may become compromised or negligible (Miller et al. 2014). Dahl et al. (2007) showed, for example, that although PGA scaffolds become fragmented and degrade to less than 15% of their initial mass within 5 weeks in culture, they lose their mechanical integrity within the first 3 weeks in culture. We captured the near-complete loss of polymer mass in PGA-P(CL/LA) scaffolds by 12 weeks using a sigmoidal decay function that models the loss of polymer mass density consistent with experimental data (Naito et al. 2013). The mechanical contribution of the polymeric scaffold was modeled with a standard neo-Hookean form (Roh et al. 2008, Niklason et al. 2010). Furthermore, based on data from prior studies (Vieira et al. 2011), polymer structural integrity was considered to decrease linearly following the observed degradation of the P(CL/LA) coating, and a parametric study was performed to identify the rate of change in the associated material parameters to match experimental data (Naito et al. 2013). Other approaches to modeling the progressive loss of structural integrity of the polymer, prior to its complete degradation, can be found elsewhere (Baek and Pence 2011, Vieira et al. 2013).

The presence of residual polymer fragments not only provides a continuing inflammatory stimulus, it may be detrimental either by creating stress concentrations or by prohibiting contiguous matrix deposition. The mechanical impact of these partially degraded remnant polymer fragments is thus not negligible. For example, densely packed PGA fragments at a suture line can act as a large circular void, forcing the collagenous tissue adjacent to the polymer to bear stresses greater than those that would be calculated given tissue homogeneity and isotropy. Therefore, not only may the residual polymer in the vessel wall not contribute to load bearing, but it may also reduce the ultimate strength of the neotissue and cause failure at loads lower than those in the absence of a polymer.

The concept of stress shielding follows directly from the scaffold's intact load-bearing capability. If synthetic cells (presumably, mainly fibroblasts and smooth muscle cells) that have infiltrated the scaffold are stress-shielded by the polymer, they likely will not initiate stress-mediated production of the extracellular matrix. Continued scaffold degradation and loss of mechanical integrity can allow a gradual transfer of the load bearing to the synthetic cells, which are mechanosensitive and should then begin to produce neotissue in response to this increased stress. Of course, if the scaffold consists of different polymeric constituents, each with a different mass fraction, density, material stiffness, and degradation rate, the contributions of each constituent (mass and load-bearing integrity) must be modeled separately in terms of both inflammatory burden and load bearing.

It is thus emphasized that although one polymer may lose mechanical integrity early on in graft development, other long-lasting polymers, even if present in comparatively smaller mass fractions, may still contribute to stress shielding of the infiltrating synthetic cells. For example, in the bilayered arterial graft developed by Wu et al. (2012), the PCL sheath models a protective "adventitia" that increases graft strength: it (initially) prevents bleeding by controlled fibrin formation within the sheath and prevents dilatation and catastrophic rupture by providing resistance to arterial blood pressures. Therefore, the outer polymeric layer supports the inner PGS core throughout its rapid degradation and replacement with the matrix. Given the much slower degradation rate of the PCL sheath, however, the stress shielding provided by the outer layer could be detrimental to stress-mediated matrix production within the inner PGS core as it degrades. It is therefore critical to select carefully the initial properties of the sheath: if it degrades too rapidly, the overall construct could rupture; if it degrades too slowly, it could limit stress-mediated matrix production while facilitating scar-like matrix production that encapsulates the construct due to a chronic foreign body response.

In summary, the mechanical and chemical properties of the scaffold, both initial and evolving, are critical to many different aspects of neovessel development (Miller et al. 2015). Given the complexity of both the evolving mechanics and the mechanobiology, the design of scaffold properties is not trivial and intuition alone cannot be expected to yield an optimal outcome.

7.3.2 Roles of Inflammation

Given that the implanted scaffold is a foreign body, there is a rapid infiltration of host inflammatory cells upon graft implantation. Roh et al. (2010) showed that the overall cellularity of TEVGs in mice increased during the first week of implantation *in vivo* and was attributable primarily to infiltration of host monocytes into the porous scaffold in response to MCP-1 production. Motivated by histological and compositional data (Hibino et al. 2011b, Naito et al. 2012, 2013), we used a gamma distribution function to model phenomenologically the inflammation-mediated production of the collagenous matrix secondary to this foreign body response (Figure 7.4a) and an exponential decay to capture the subsequent degradation of this collagen (Miller et al. 2014). Coinciding with the experimental data, the simulated inflammation-mediated synthesis of collagen peaks at the time of the highest monocyte infiltration (10 days) (Hibino et al. 2011) and ceases upon complete polymer degradation (Naito et al. 2013). Given the lack of complete information, it was assumed that the synthesis and deposition of inflammation-mediated collagen was isotropic, producing similar fold increases for all four collagen fiber families in response to the implanted scaffold. Although it is not known how much of the collagen that is synthesized via a foreign body response is lost with the degradation of the polymer, its half-life was modeled to decrease in the presence of increased matrix metalloproteinase (MMP) expression (Strauss et al. 1996), which occurs during the early remodeling process as a consequence of monocyte responses. The inflammation-mediated collagen production is assumed to have microstructural and compositional properties that render it stiffer than the native collagen normally present in the murine abdominal IVC (Lee et al. 2013), that is, larger fiber diameters, a higher type I:III ratio, increased cross-linking, and so on (Khosravi et al. 2015). It should be noted that if the pore size of the scaffold does not allow inflammatory cell infiltration, the foreign body response still drives the deposition of this inflammation-mediated collagen production, albeit in the form of fibrous encapsulation of the construct (Sanders et al. 2000, Junge et al. 2012).

The structural and material properties of the scaffold, not just its surface chemistry, also play important roles in modulating the host inflammatory response (Sanders et al. 2005, Boehler et al. 2011, van Loon et al. 2013). Numerous experimental studies of structure–function relations demonstrate the effects of pore size, porosity, polymer stiffness, degradation rate, fiber diameter, and fiber alignment (to name a few) on inflammatory cell recruitment and activation (Miller et al. 2015). For example, it is well established that pore size is a primary determinant of successful inflammatory as well as synthetic cell infiltration (Balguid et al. 2009, Choi et al. 2010, Boccafoschi et al. 2012, van Loon et al. 2013, Zander et al. 2013). If cells cannot invade the polymer, a collagenous scar plate will encapsulate the scaffold, which increases the probability of fibrosis or graft failure (Zander et al. 2013). It has been suggested that 10 µm is the minimum pore size for cells to infiltrate a scaffold without detrimental shearing, though the optimal value remains unclear and may differ across cell types (Alberts et al. 2002, Eichhorn

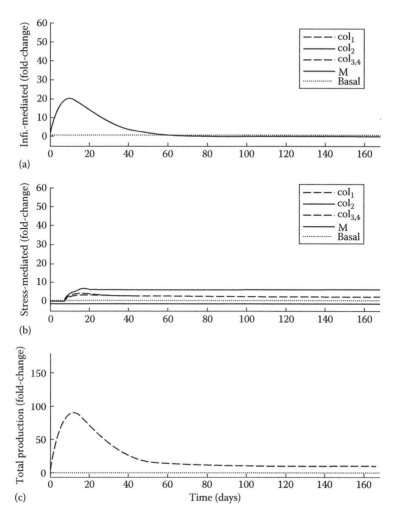

Figure 7.4 Mass density production for each constituent (collagen families in the axial (1), circumferential (2), and diagonal (3 and 4) directions, and circumferential smooth muscle) throughout the growth and remodeling simulation. (a) Inflammation-mediated production peaks at 10 days (reflecting the time of the highest measured monocyte/macrophage infiltration) and subsequently decreases following loss of the polymeric scaffold. (b) Stress-mediated production increases following the loss of scaffold structural integrity: increases are anisotropic and remained elevated relative to basal for the remainder of the simulation. (c) Total increases in production for all constituents relative to basal rates suggest a steady state, although long-term turnover is suboptimal to that of the native IVC. (Modified from Miller, K.S. et al., *Acta Biomater.*, 11, 283, 2015.)

and Sampson 2005). Scaffold fiber diameter can influence cell phenotype and differentiation as it determines the surface area on which cells can attach and spread (Sanders et al. 2000, 2005). Increasing fiber diameter increases the magnitude of the inflammatory response (through macrophage activation and secretion of proinflammatory molecules) and subsequent matrix accumulation

(Lam et al. 1995, Sanders et al. 2002, Boland et al. 2004, Garg et al. 2013). Fiber diameters below 6 μm (if separated adequately) seem to reduce the number of activated macrophages (cells remain quiescent) and fibrous capsule thickness (Sanders et al. 2000, 2005). Correspondingly, increased porosity also increases the surface area available for cell infiltration, attachment, and foreign body detection (Lam et al. 1995). Porosity is, of course, a major determinant of polymer apparent density and scaffold mechanical properties as well (Gibson and Ashby 1988). The search for an optimal design must account for potentially competing effects. Importantly, the degradation rate for a given polymer depends strongly on both porosity and fiber diameter, which also determine the available surface area for polymer–cell interactions (Lam et al. 1995). Finally, scaffold alignment not only influences the mechanical properties of the implanted construct (in particular, the magnitude of its anisotropy), it can also guide cell and matrix alignment while the fibers are present (Mauck et al. 2009, Cao et al. 2010, Milleret et al. 2011). It is thought that a transition occurs upon fiber degradation, with cell and matrix alignment then dictated instead by the mechanical stresses or strains experienced by synthetic cells.

7.3.3 Simulated Neovessel Formation

Recall that there will be no stress-mediated matrix production while synthetic cells are stress-shielded; as the polymer degrades and its structural integrity becomes compromised, however, these cells begin to sense increasing loads and to deposit the matrix accordingly to minimize deviations in intramural stress from the preferred homeostatic value (Figure 7.4b). Given that the scaffold is unseeded upon implantation, we accounted for the time required for cell infiltration, as collagen production cannot begin until host synthetic cells have invaded the construct (Miller et al. 2014). Finally, it was assumed in the long term the graft will reach a basal rate of collagen production (Figure 7.4c). In contrast to the isotropic production of inflammation-mediated collagen, the fold increases in stress-mediated production (relative to basal values) were anisotropic (in other words, different for different families of collagen fibers) because of variations in tensions from homeostatic values in different directions. This anisotropy allows an appropriate evolution of the organization of collagen in response to the applied loads.

In accounting for the effects of both the early inflammatory phase and subsequent mechano-mediated phase on the kinetics of matrix production, this model of *in vivo* neovessel development was able to capture the mechanical behavior of the evolving graft. We found a transition from the stiff PGA-P(CL/LA) scaffold (neo-Hookean) augmented by stiff collagen (higher type I:III ratio) produced via the inflammatory response at 2 weeks to a more compliant (Fung-type exponential) response from 6 to 24 weeks (Figure 7.3). This evolving response reflects both the decreasing contribution of the polymeric scaffold and the transition from inflammation- to stress-mediated collagen production (which begins approximately 4–5 weeks after implantation when the scaffold loses integrity). The behavior at 6–12 weeks is representative of the coexistence of collagen produced via both mechanisms. At 24 weeks, the graft consists primarily of collagen produced via the stress-mediated mechanism and is therefore more compliant. The predicted evolving material (wall tension–stretch relationship) and structural (pressure-diameter behavior) responses were similar to those observed in experimental biaxial tests (Figure 7.3) (Naito et al. 2013), suggesting that the computational model can capture salient aspects of collagen deposition throughout neovessel development.

7.4 PROMISE OF COMPUTATIONAL MODELS

Recent advances in the tissue engineering of vascular grafts have enabled a paradigm shift from the desire to design for adequate burst pressure, suture retention, and thrombo-resistance to the goal of achieving grafts having properties of healthy native vessels, including long-term mechanobiological stability and growth potential (Roh et al. 2008, Naito et al. 2013, Khosravi et al. 2015). Compliance mismatch between a vascular graft and the native vessel has long been considered a critical determinant of graft failure, leading at times to the formation of neointimal hyperplasia and stenosis (Seifu et al. 2013). Excessive TEVG stiffness should therefore be avoided, as it can lead to compliance mismatch and thus altered hemodynamics that induce unfavorable mechanobiological consequences (Miller et al. 2015). Achieving this far more ambitious outcome requires the identification of optimal, not just adequate, scaffold structure and material properties to yield the desired *in vivo* functionality. Given the almost limitless combinations of scaffold parameters, there is a pressing need to move beyond purely empirical trial-and-error approaches and toward rational design of vascular grafts. Hypothesis-driven computational models can enable time- and cost-efficient evaluations of fundamental hypotheses (Valentin and Humphrey 2009) and parameter sensitivity studies (Sankaran et al. 2013), and thereby can reduce the experimental search space. Such models allow us to assess *in silico* the long-term consequences of different combinations of scaffold structural and material properties via parametric studies, and to identify scaffold designs believed to be most promising in terms of *in vivo* functionality for subsequent experimental validation (Miller et al. 2015).

Given that there are many structural and material parameters of polymeric scaffolds that can influence both the mechanical performance of an implanted graft and its biological function through modulation of the host inflammatory response (cf. van Loon et al. 2013), we initially reduced this extensive list to six parameters most frequently investigated experimentally and physically tractable (Miller et al. 2015). Using the Buckingham Pi approach to accomplish

Table 7.2: Buckingham Pi Nondimensional Analysis Performed by Miller et al. (2015) to Reduce Tractable Physical Parameters for a Fibrous Scaffold from 6 to 4

Inflammatory Response, $y = f\left(k_q^p, \varepsilon, c^p, \omega, r, \phi^k\right)$

Scaffold Parameter	Symbol	SI Units	General Units	π Group
Degradation rate	k_q^p	days^{-1}	$L^0T^{-1}M^0$	1
Porosity	ε	—	$L^0T^0M^0$	ε
Modulus	c^p	MPa	$L^{-1}T^{-2}M^1$	1
Fiber diameter	ω	cm	$L^1T^0M^0$	ω/r_{min}
Pore size	r	cm	$L^1T^0M^0$	r/r_{min}
Alignment	ϕ^k	—	$L^0T^0M^0$	ϕ^k

Scales for L (Length), T (Time), and M (Mass) — **Inflammatory Response**

$$L_s = r_{min}, T_s = 1/k_q^p, M_s = c^p r_{min}/\left(k_q^p\right)^2$$

$$y = \breve{f}\left(\varepsilon, \omega/r_{min}, r/r_{min}, \phi^k\right)$$

Source: Modified from Miller, K.S. et al., *Acta Biomater.*, 11, 283, 2015.
Notes: L, T, and M denote length, time, and mass. Scales for the three primary dimensions, or units, are also listed, along with a reduced list of nondimensional parameters.

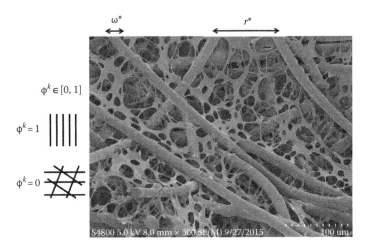

Figure 7.5 Scanning electron microscopy image of a PGA-P(CL/LA) scaffold showing the primary parameters considered computationally: normalized fiber diameter ω and normalized pore size r. Scaffold alignment should also be considered in the future when experimental data become available, where $\phi^k = 1$ indicates a highly aligned scaffold and $\phi^k = 0$ a scaffold with randomly organized fibers. (SEM image: Courtesy of Cameron Best, Nationwide Children's Hospital, Columbus, OH; Modified from Miller, K.S. et al., *Acta Biomater.*, 11, 283, 2015.)

non-dimensionalization (cf. Humphrey and O'Rourke 2015) (Table 7.2), we considered a functional response in terms of three primary nondimensional parameters of interest: porosity, fiber diameter, and alignment, with pore size being expressed as a function of porosity and fiber diameter, modulus as a function of alignment and porosity, and degradation rate strongly influenced by porosity and fiber diameter, which determine the available surface area for polymer–cell interactions (Figure 7.5). We then used observed structure–function relationships from the literature, including postulated mechanical and biological consequences, to motivate illustrative constitutive relations for each of these three nondimensional scaffold parameters (see Table 1 in Miller et al. (2015) and references therein). By simulating the evolution of the linearized material stiffness (which dictates hemodynamic consequences) throughout graft development and comparing it with the target value for healthy, normal murine IVCs, we identified initial groups of preferred combinations of scaffold physical parameters that could reduce excessive stiffness throughout graft evolution and thus avoid compliance mismatch between the graft and the native IVC, which should help to prevent potentially unfavorable mechanobiological consequences throughout long-term implantation. Given that the model simulates not just an initial response but also a long-term evolutionary process for each potential scaffold design, there remains a need to perform parameter sensitivity studies, to use an efficient approach for uncertainty quantification, and to achieve optimization by minimizing cost functions that contrast the evolving biaxial material stiffness and the associated structural stiffness of the TEVG and the host vessel (cf. Sankaran et al. 2013).

7.4.1 Biochemomechanical Motivations for Constitutive Relations
Advances in fabrication techniques such as electrospinning now enable one to carefully fine-tune key scaffold parameters—as a result, there is a need to

understand better how the structure and material properties of the polymeric construct influence neovessel remodeling. Toward this end, we extended our earlier model to consider the influence of scaffold physical parameters on the foreign body response during the inflammatory period of neotissue development (Miller et al. 2015). It was conjectured that neotissue development (matrix production and removal) is driven largely by pore size, noting that pore size relates directly to porosity and polymer fiber diameter (Eichhorn and Sampson 2005). New data-driven constitutive relations were used to predict matrix kinetics as a function of scaffold parameters as well as deviations in mechanical stress from a preferred homeostatic state for each matrix constituent (Valentin and Humphrey 2009, Miller et al. 2015).

The results from the literature on observed structure–function relationships motivated the selection of constitutive relations, including postulated mechanical and biological consequences for each parameter (Table 1 in Miller et al. 2015). For example, it was conjectured that interactions between infiltrating monocytes and synthetic cells could be modeled phenomenologically as a function of the mean pore size (Balguid et al. 2009, Boccafoschi et al. 2012, van Loon et al. 2013, Zander et al. 2013), which yielded simulations that matched well with the available macroscopic data. The corresponding functional form was motivated by qualitative information that increased fiber diameter and increased porosity (which coincides with increased pore size) stimulate greater inflammatory responses and subsequent accumulation of the matrix (Lam et al. 1995, Sanders et al. 2002, Boland et al. 2004, Balguid et al. 2009, Boccafoschi et al. 2012, Garg et al. 2013, van Loon et al. 2013, Zander et al. 2013). It was also postulated that because the inflammatory response increases cytokines, chemokines, and growth factors in the extracellular milieu, it may increase both the proliferative capability of intramural synthetic cells as well as their ability to respond to deviations from the preferred state of stress. In this way, potentially coupled inflammatory and mechano-mechanisms can be captured phenomenologically. Correspondingly, it was assumed that increased inflammation involves increased macrophages and neutrophils, and thus MMPs, which may decrease the constituent half-life (Strauss et al. 1996).

Indeed, the simulated results captured the hypotheses and yielded information on emergent behaviors. For example, for a fixed porosity, decreasing the fiber diameter decreased the inflammatory response, which in turn decreased matrix production and indirectly increased the compliance of the TEVG at later times (Figure 7.6a) (Miller et al. 2015). Simulated tension–stretch relationships also demonstrated that, for a fixed fiber diameter, decreasing the porosity increased the initial mechanical contribution of the polymer; however, following polymer degradation, smaller values of porosity yielded less of an inflammatory response, hence less matrix production and a more compliant TEVG at later times (Figure 7.6b) (Miller et al. 2015).

Importantly, this model suggests that certain combinations of physical parameters may result in excessive stiffness and large compliance mismatches between the TEVG and adjacent vein. High values of stiffness may result in an altered cell phenotype, which may increase the risk of chronic inflammation and fibrosis (Fereol et al. 2006, Chan and Mooney 2008, Blakney et al. 2012, Forte et al. 2012). Large fluctuations in stiffness throughout TEVG evolution may also result in unstable or unpredictable cell phenotypes, which may impede one's ability to evaluate or predict patient progress clinically (Sussman et al. 2013, Wang et al. 2014). Thus, we suggested that such combinations should be avoided to prevent potentially unfavorable or unpredictable mechanobiological consequences (Fereol et al. 2006, Chan and Mooney 2008, Blakney et al. 2012, Forte et al. 2012). The results from parametric studies yet revealed

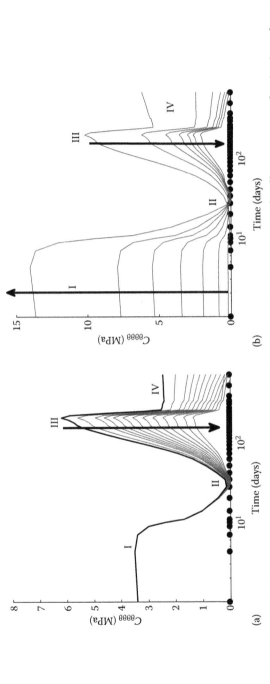

Figure 7.6 Effect of fiber diameter ω and porosity on stiffness. Values of circumferential linearized stiffness, computed at *in vivo* values of circumferential and axial stretch, are contrasted between the simulated TEVGs (solid lines) and native IVC (solid circles). (a) The solid black line shows results for the implanted scaffold, with an initial normalized fiber diameter ω = 1.4 and porosity ε = 80%; the solid gray lines show results for ω ∈ [0.01, 1.4], also for ε = 80%. Linearized stiffness was initially dominated by the mechanical properties of the polymeric scaffold (phase I), with a subsequent decrease due to the polymer losing its load-bearing integrity (phase II), a late peak caused by accumulation of ECM from overlapping inflammation- and mechano-mediated deposition (phase III) and final plateau from mechano-mediated turnover in a steady state (phase IV). Decreasing ω (indicated by the direction of the dark arrow) indirectly decreased matrix production and subsequently resulted in a lower peak (III) and plateau (IV) linearized stiffness. (b) The gray lines show results for initial scaffold porosity ε ∈ [60%, 95%] and a representative normalized fiber diameter ω* = 0.6. Decreasing the porosity (indicated by the direction of the dark arrow) increased the initial contribution of the scaffold (phase I) and indirectly decreased matrix production, resulting in decreased peak (III) and plateau values (IV) of linearized stiffness. The model suggested that ε ∈ [75%, 85%] decreased the overall variation in stiffness as the TEVG evolved and should therefore be explored experimentally. (Reprinted with permission of Miller, K.S. et al., *Acta Biomater.*, 11, 283, 2015.)

that several sets of physical parameters may reduce deviations in linearized stiffness, with overall compliance approaching that of the native IVC at 2 years, and should be considered when designing future scaffolds for experimental validation (Miller et al. 2015). Despite the encouraging results achieved thus far, there still remains a pressing need for significantly more longitudinal data to inform, validate, and extend this and similar models.

7.4.2 Modeling Phenotypic Diversity

To demonstrate the predictive capability of the G&R computational model and to evaluate the long-term mechanobiological stability of PGA-P(CL/LA) scaffolds implanted in the murine venous circulation, we next extended simulations out to 2 years (essentially, the lifespan of the mouse) and evaluated the biaxial mechanical behavior of TEVGs explanted at this time point (Khosravi et al. 2015). Remarkably, the previously identified set of model parameters that fit the mean TEVG data until 24 weeks post-implantation (Miller et al. 2014) similarly predicted the mean response up to ~100 weeks (i.e., 2 years) without any modifications. Once it was established that the model captured salient features of long-term *in vivo* neovessel development, we then explored how the model could test hypotheses underlying the observation of biomechanical diversity.

Minor variations in graft physical properties may have far-reaching effects on local degradation characteristics of the polymer and its load-bearing integrity, which in turn could change the mechanical stimuli experienced by the resident cells and thereby influence local cell phenotype and matrix production (Choi et al. 2010). In particular, although the mean biaxial mechanical behavior of the neovessels approached that of the native IVC, significant biomechanical diversity manifested in terms of graft stiffness, particularly after 24 weeks of implantation. The computational model suggested that the observed phenotypic diversity likely resulted from differences in evolving collagen properties, which suggested the need for particular *a posteriori* experimental assays that indeed confirmed the model predictions.

Following a series of parametric studies to determine which model parameters could delineate the most versus the least distensible neovessels, we found that the long-term development of stiff neovessels could be simulated computationally by increasing the material stiffness of the collagen that was produced during the mechano-mediated period. This phenomenological modification of collagen parameters was motivated by possible differences in collagen composition such as an increased collagen type I:III ratio or an increased number of collagen cross-links. Conversely, the development of compliant neovessels could be simulated by decreasing the material stiffness of the collagen produced during the mechano-mediated period (i.e., a decrease in type I:III collagen ratio or decrease in cross-linking) or by decreasing the half-life of the collagen produced during the inflammatory period in response to polymer degradation.

7.4.3 Uncertainty Quantification, Sensitivity Analysis, and Optimization

When modeling the long-term evolution of TEVGs in the venous or arterial circulation for multiple initial scaffold designs, it becomes necessary to use an efficient approach for optimization. Sankaran et al. (2013) discuss at length the choice of methods for parameter sensitivity, uncertainty quantification, and optimization in G&R simulations. Here, a brief summary of this approach is provided in the context of vascular grafts.

Consider, for example, a study based on the model presented earlier. First, one could perform a series of computationally efficient parameter sensitivity studies focusing on the four primary classes of parameters: scaffold and matrix stiffness, scaffold geometry (e.g., porosity, fiber diameter, and pore size), scaffold and matrix degradation rates, and matrix production rates (inflammation- and mechano-mediated). This overall process should include systematic and efficient sensitivity analyses and uncertainty quantification, which involves multiple steps: generating an initial set of collocation points, running simulations at each point, and computing statistics in terms of a probability distribution function or confidence interval. The level of refinement of the collocation grid can be increased hierarchically to achieve a convergence of the statistics (Sankaran and Marsden 2011). When such an approach was used to assess a standard G&R analysis of arterial adaptation, the results were compared directly with those achieved based on fitting experimental data using a "brute force" study of parameter sensitivity (Valentin and Humphrey 2009). The new method revealed similar findings and yet was much more efficient than the "trial-and-error" approach (Ramachandra et al. 2015). By utilizing a systematic and efficient framework to evaluate the sensitivity and subsequent uncertainty of a G&R model of neovessel formation, one should be able to identify key parameters of interest to inform the next set of experiments. In particular, uncertainty quantification may identify the class of parameters (recall: scaffold and matrix stiffness, scaffold geometry, scaffold and matrix degradation rates, and matrix production rates) that the system is most sensitive to, hence potentially yielding those with the most promise for improving current clinical outcomes.

Next, parameter optimization can be performed using a Surrogate Management Framework derivative-free approach, which has a well-established convergence theory (Sankaran et al. 2013). An optimal parameter set that defines the best scaffold design(s) can be determined by minimizing an appropriate objective function that includes statistical moments. One goal could be to minimize differences in biaxial *in vivo* material stiffness between the evolving TEVG and the adjacent host vessel; in other words, one mechanobiological target for optimization could be the biaxial material stiffness to avoid compliance mismatch throughout the implantation period (Seifu et al. 2013). By coupling G&R simulations with the aforementioned adaptive sparse grid collocation approach and derivative-free optimization algorithm, it is hoped that one could optimize scaffold parameters wherein the evolving neovessel can maintain homeostatic conditions (e.g., biaxial material stiffness) over physiologic perturbations in pressure and flow (cf. Sankaran et al. 2013). Identifying scaffold design(s) that permit neovessel adaptation to evolving mechanical loading (growth potential) will be vital to the success of TEVGs, particularly those utilized in the repair of congenital heart defects, which must adapt to the growing child following implantation. Herein, we submit that uncertainty quantification, sensitivity analysis, and robust optimization coupled with computational G&R models could be valuable tools to identify optimal scaffold design(s), evaluate neovessel growth potential, and move closer toward the rational design of TEVGs.

7.5 CONCLUSIONS

Computational biomechanical models show great promise in advancing the field of vascular tissue engineering. They enable time- and cost-efficient evaluations of fundamental hypotheses to reduce the experimental search space and aid investigators in moving away from a purely empirical trial-and-error approach and toward rational design. This approach enables hypothesis-driven studies of the cellular and molecular mechanisms of neovessel formation, factors influencing the inflammatory response and cellular phenotypes, and the causes and consequences of biomechanical diversity. Model predictions can, in turn, be used to guide subsequent experimental validation, and suggested sets of model parameters can be used in parameter sensitivity studies, uncertainty quantification, and optimization. Finally, computational models can be extended to include additional mechanobiological information as it becomes available, including the consequences of cross talk between mural and inflammatory cells, interactions between inflammatory and mechanical responses, roles of scaffold alignment and other modifiable parameters, polymer degradation profiles, and, most significantly, in the arterial circulation, dual-layer scaffold designs.

ACKNOWLEDGMENT

This work was supported, in part, by grants from the NIH (R01 EB008836, R01 HL098228, and R01 HL128602).

REFERENCES

Alberts, B., Johnson, A., Lewis, J., Raff, M., Roberts, K., and Walter, P. 2002. *Molecular Biology of the Cell*, 4th edn. New York: Garland Science.

American Heart Association Statistics Committee and Stroke Statistics Subcommittee. 2015. Heart disease and stroke statistics—2015 update: A report from the American Heart Association. *Circulation* 131:e29–e322.

Baek, S. and Pence, T. 2011. On mechanically induced degradation of fiber-reinforced hyperelastic materials. *Math Mech Solid* 16:406–434.

Balguid, A., Mol, A., van Marion, M.H., Bank, R.A., Bouten, C.V.C., and Baaijens, F.P.T. 2009. Tailoring fiber diameter in electrospun scaffold for optimal cellular infiltration in cardiovascular tissue engineering. *Tissue Eng A* 15:437–444.

Blakney, A.K., Swartzlander, M.D., and Bryant, S.J. 2012. The effects of substrate stiffness on the in vivo activation of macrophages and in vivo host response to poly(ethylene glycol)-based hydrogels. *J Biomed Mater Res A* 100A:1375–1386.

Boccafoschi, F., Mosca, C., and Cannas, M. 2012. Cardiovascular biomaterials: When the inflammatory response helps to efficiently restore tissue functionality? *J Tissue Eng Regen Med* 8:253–267.

Boehler, R.M., Graham, J.G., and Shea, L.D. 2011. Tissue engineering tools for modulation of the immune response. *Biotechniques* 51:239–253.

Boland, E.D., Telemeco, T.A., Simpson, D.G., Wnek, G.E., and Bowlin, G.L. 2004. Utilizing acid pretreatment and electrospinning to improve biocompatibility of poly(glycolic acid) for tissue engineering. *J Biomed Mater Res B* 71B:144–152.

Cao, H., McHugh, K., Chew, S.Y., and Anderson, J.M. 2010. The topographical effect of electrospun nanofibrous scaffolds on the in vivo and in vitro foreign body reaction. *J Biomed Mater Res A* 93:1151–1159.

Chan, G. and Mooney, D.J. 2008. New materials for tissue engineering: Towards greater control over the biological response. *Trends Biotechnol* 26:382–392.

Choi, S.W., Zhang, Y., and Xia, Y. 2010. Three-dimensional scaffolds for tissue engineering: The importance of uniformity in pore size and structure. *Langmuir* 26:19001–19006.

Dahl, S.L., Kypson, A.P., Lawson, J.H., Blum, J.L., Strader, J.T., Li, Y., Manson, R.J. et al. 2011. Readily available tissue-engineered vascular grafts. *Sci Transl Med* 368:68ra9.

Dahl, S.L.M., Rhim, C., Song, Y.C., and Niklason, L.E. 2007. Mechanical properties and compositions of tissue engineered and native arteries. *Ann Biomed Eng* 35:348–355.

Eichhorn, S.J. and Sampson, W.W. 2005. Statistical geometry of pores and statistics of porous nanofibrous assemblies. *J R Soc Interface* 2:309–318.

Fereol, S., Fodil, R., Labat, B., Galiacy, S., Laurent, V.M., Louis, B., Isabey, D., and Planus, E. 2006. Sensitivity of alveolar macrophages to substrate mechanical and adhesive properties. *Cell Motil Cytoskeleton* 63:321–340.

Forte, G., Pagliara, S., Ebara, M., Uto, K., Tam, J.K.V., Romanazzo, S., Escobedo-Lucea, C., Romano, E., Nardo, P.D., Traversa, E., and Aoyagi, T. 2012. Substrate stiffness modulates gene expression and phenotype in neonatal cardiomyocytes in vitro. *Tissue Eng A* 18:1837–1848.

Garg, K., Pullen, N.A., Ozkeritzian, C.A., Ryan, J.J., and Bowlin, G.L. 2013. Macrophage functional polarization (M1/M2) in response to varying fiber and pore dimensions of electrospun scaffolds. *Biomaterials* 34:4439–4451.

Gibson, L.J. and Ashby, M.F. 1988. *Cellular Solids: Structure and Properties.* Oxford, U.K.: Pergamon Press.

Gleason, R.L., Gray, S.P., Wilson, E., and Humphrey, J.D. 2004. A multiaxial computer-controlled organ culture and biomechanical device for mouse carotid arteries. *ASME J Biomech Eng* 126:787–795.

Hibino, N., McGillicuddy, E., Matsumura, G., Ichihara, Y., Naito, Y., Breuer, C., and Shinoka, T. 2010. Late-term results of tissue engineered vascular grafts in humans. *J Thorac Cardiovasc Surg* 139:431–436.

Hibino, N., Villalona, G., Pietris, N., Duncan, D.R., Schoffner, A., Roh, J.D., Yi, T. et al. 2011a. Tissue-engineered vascular grafts form neovessels that arise from regeneration of the adjacent blood vessel. *FASEB J* 25:2731–2739.

Hibino, N., Yi, T., Duncan, D.R., Rathore, A., Dean, E., Naito, Y., Dardik, A. et al. 2011b. A critical role for macrophages in neovessel formation and the development of stenosis in tissue-engineered vascular grafts. *FASEB J* 25:4253–4263.

Hu, J.J. 2015. Constitutive modeling of an electrospun tubular scaffold used for vascular tissue engineering. *Biomech Model Mechanobiol* 14:897–913.

Humphrey, J.D. and O'Rourke, S.L. 2015. *An Introduction to Biomechanics: Solids and Fluids, Analysis and Design*, 2nd edn., New York: Springer.

Humphrey, J.D. 2002. *Cardiovascular Solid Mechanics: Cells, Tissues, and Organs.* New York: Springer.

Humphrey, J.D. and Rajagopal, K.R. 2002. A constrained mixture model for growth and remodeling of soft tissues. *Math Models Methods Appl Sci* 12:407.

Jeffries, E.M., Allen, R.A., Gao, J., Pesce, M., and Wang, Y. 2015. Highly elastic and suturable electrospun poly(glycerol sebacate) fibrous scaffolds. *Acta Biomater* 18:30–39.

Junge, K., Binnebosel, M., von Trotha, K.T., Rosch, R., Klinge, U., Neumann, U.P., and Jansen, P.L. 2012. Mesh biocompatibility: Effects of cellular inflammation and tissue remodeling. *Langenbecks Arch Surg* 397:255–270.

Kannan, R.Y., Salacinski, H.J., Butler, P.E., Hamilton, G., and Seifalian, A.M. 2005. Current status of prosthetic bypass grafts: A review. *J Biomed Mater Res B Appl Biomater* 74:570–581.

Khosravi, R., Best, C.A., Allen, R.A., Stowell, C.E., Onwuka, E., Zhuang, J.J., Lee, Y.U. et al. 2016. Long-term functional efficacy of a novel electrospun poly(glycerol sebacate)-base arterial graft in mice. *Ann Biomed Eng* 44(8):2402–2416.

Khosravi, R., Miller, K.S., Best, C.A., Shih, Y.C., Lee, Y.U., Yi, T., Shinoka, T., Breuer, C.K., and Humphrey, J.D. 2015. Biomechanical diversity despite mechano-biological stability in tissue engineered vascular grafts two years post-implantation. *Tissue Eng A* 21:1529–1538.

Lam, K.H., Schakenraad, J.M., Groen, H., Esselbrugge, H., Dijkstra, P.J., Feijen, J., and Nieuwenhuis, P. 1995. The influence of surface morphology and wettability on the inflammatory response against PLLA: A semi-quantitative study with monoclonal antibodies. *J Biomed Mater Res* 29:929–942.

Lee, Y.U., Naito, Y., Kurobe, H., Breuer, C.K., and Humphrey, J.D. 2013. Biaxial mechanical properties of the inferior vena cava in C57BL/6 and C-17 SCID/bg mice. *J Biomech* 46:2277–2282.

L'Heureux, N., Dusserre, N., Konig, G., Victor, B. Keire, P., Wight, T.N., Chronos, N.A. et al. 2006. Human tissue-engineered blood vessels for adult arterial revascularization. *Nat Med* 12:361–365.

Mauck, R.L., Baker, B.M., Nerurkar, N.L., Burdick, J.A., Li, W.L., Tuan, R.S., and Elliott, D.M. 2009. Engineering on the straight and narrow: The mechanics of nanofibrous assemblies for fiber-reinforced tissue regeneration. *Tissue Eng B* 15:171–193.

Miller, K.S., Khosravi, R., Breuer, C.K., and Humphrey, J.D. 2015. A hypothesis-driven parametric study of effects of polymeric scaffold properties on tissue engineered neovessel formation. *Acta Biomater* 11:283–294.

Miller, K.S., Lee, Y.U., Naito, Y., Breuer, C.K., and Humphrey, J.D. 2014. Computational model of in vivo neovessel development from an engineered polymeric vascular construct. *J Biomech* 47:2080–2087.

Milleret, V., Simona, B., Neuenschwander, P., and Hall, H. 2011. Tuning electrospinning parameters for production of 3D fiber fleeces with increased porosity for soft tissue engineering applications. *Euro Cells Mater* 21:286–303.

Naito, Y., Lee, Y.U., Yi, T., Church, S.N., Solomon, D., Humphrey, J.D., Shinoka, T., and Breuer, C.K. 2013. Beyond burst pressure: Initial evaluation of the natural history of biaxial mechanical properties of tissue-engineered vascular grafts in the venous circulation using a murine model. *Tissue Eng A* 20:346–355.

Naito, Y., Williams-Fritze, M., Duncan, D.R., Church, S.N., Hibino, N., Madri, J.A., Humphrey, J.D., Shinoka, T., and Breuer, C.K. 2012. Characterization of the natural history of extracellular matrix production in tissue-engineered vascular grafts during neo-vessel formation. *Cells Tissues Organs* 195:60–72.

Niklason, L.E., Gao, J., Abbott, W.M., Hirschi, K.K., Houser, S., Marini, R., and Langer, R. 1999. Function arteries grown in vitro. *Science* 284:489–493.

Niklason, L.E., Yeh, A.T., Calle, E., Bai, Y., Valentin, A., and Humphrey, J.D. 2010. Enabling tools for engineering collagenous tissues, integrating bioreactors, intravital imaging, and biomechanical modeling. *Proc Natl Acad Sci* 107:3335–3339.

Patterson, J.T., Gilliland, T., Maxfield, M.W., Church, S., Naito, Y., Shinoka, T., and Breuer, C.K. 2012. Tissue-engineered vascular grafts for use in the treatment of congenital heart disease: From the bench to the clinic and back again. *Regen Med* 7:409–419.

Ramachandra, A.B., Sankaran, S., Humphrey, J.D., and Marsden, A.L. 2015. Computational simulation of the adaptive capacity of vein grafts in response to increased pressure. *ASME J Biomech Eng* 137(3):031009.

Roh, J.D., Nelson, G.N., Brennan, M.P., Mirensky, T.L., Yi, T., Hazlett, T.F., Tellides, G. et al. 2008. Small-diameter biodegradable scaffolds for functional vascular tissue engineering in the mouse model. *Biomaterials* 29:1454–1463.

Roh, J.D., Sawh-Martinez, R., Brennan, M.P., Jay, S.M., Devine, L., Rao, D.A., Yi, T. et al. 2010. Tissue-engineered vascular grafts transform into mature blood vessels via an inflammation-mediated process of vascular remodeling. *Proc Natl Acad Sci* 107:4669–4674.

Sanders, J.E., Cassisi, D.V., Neumann, T., Golledge, S.L., Zachariah, S.G., Ratner, B.D., and Bale, S.D. 2002. Relative influence of polymer fiber diameter and surface charge on fibrous capsule thickness and vessel density for single-fiber implants. *J Biomed Mater Res A* 65A:462–467.

Sanders, J.E., Lamont, S.E., Mitchell, S.B., and Malcom, S.G. 2005. Small fiber diameter fibro-porous meshes: Tissue response sensitivity to fiber spacing. *J Biomed Mater Res A* 72:335–342.

Sanders, J.E., Stiles, C.E., and Hayes, C.L. 2000. Tissue response to single-polymer fibers of varying diameters: Evaluation of fibrous encapsulation and macrophage density. *J Biomed Mater Res A* 52:231–237.

Sankaran, S., Humphrey, J.D., and Marsden, A.L. 2013. An efficient framework for optimization and parameter sensitivity analysis in arterial growth and remodeling computations. *Comput Meth Appl Mech Eng* 256:200–212.

Sankaran, S. and Marsden, A.L. 2011. A stochastic collocation method for uncertainity quantification in cardiovascular simulations. *ASME J Biomech Eng* 133:031001.

Seifu, D.G., Purnama, A., Mequanint, K., and Mantovani, D. 2013. Small-diameter vascular tissue engineering. *Nat Rev Cardiol* 10:410–421.

Sokolis, D.P. 2012. Experimental investigation and constitutive modeling of the 3D histomechanical properties of vein tissue. *Biomech Model Mechanobiol* 12:431–451.

Strauss, B.H., Robinson, R., Batchelor, W.B., Chisholm, R.J., Ravi, G., Natarajan, M.K., Logan, R.A. et al. 1996. In vivo collagen turnover following experimental balloon angioplasty injury and the role of matrix metalloproteinases. *Circ Res* 79:541–550.

Sussman, E.M., Halpin, M.C., Muster, J., Moon, R.T., and Ratner B.D. 2013. Porous implants modulate healing and induce shifts in local macrophage polarization in the foreign body reaction. *Ann Biomed Eng* 42:1508–1516.

Tamini, E., Ardila, D.C., Haskett, D.G., Doetschman, T., Slepian, M.J., Kellar, R.S., and Vande Geest, J.P. 2016. Biomechanical comparison of glutaraldehyde-crosslinked gelatin fibrinogen electrospun scaffolds to porcine coronary arteries. *ASME J Biomech Eng* 138(1):011001.

Udelsman, B.V., Khosravi, R., Miller, K.S., Dean, E.W., Bersi, M.R., Rocco, K., Yi, T., Humphrey, J.D., and Breuer, C.K. 2014. Characterization of evolving biomechanical properties of tissue-engineered vascular grafts in the arterial circulation. *J Biomech* 47:2070–2079.

Valentin, A. and Humphrey, J.D. 2009. Evaluation of fundamental hypotheses underlying constrained mixture models of arterial growth and remodeling. *Phil Trans R Soc A* 367:3585–3606.

van Loon, S.L.M., Smits, A.I.P.M., Driessen-Mol, A., Baaijens, F.P.T., and Bouten, C.V.C. 2013. The immune response in in situ tissue engineering of aortic heart valves. *Calcific Aortic Valve Disease*, Aikawa, E. (Ed.). InTech, DOI: 10.5772/54354. Available from: http://www.intechopen.com/books/calcific-aortic-valve-disease/the-immune-response-in-in-situ-tissue-engineering-of-aortic-heart-valves.

Vieira, A.C., Guedes, R.M., and Tita, V. 2013. Considerations for the design of polymeric biodegradable products. *J Polym Eng* 33:293–302.

Vieira, A.C., Vieira, J.C., Ferra, J.M., Magalhaes, F.D., Guedes, R.M., and Marques, A.T. 2011. Mechanical study of PLA-PCL fibers during in vitro degradation. *J Mech Behav Biomed Mat* 4:451–460.

Wang, Z., Cui, Y., Wang, J., Yang, X., Wu, Y., Wang, K., Gao, X. et al. 2014. The effect of thick fibers and large pores of electrospun poly(ε-caprolactone) vascular grafts on macrophage polarization and arterial regeneration. *Biomaterials* 35:5700–5710.

Wu, W., Allen, R.A., and Wang, Y. 2012. Fast-degrading elastomer enables rapid remodeling of a cell-free synthetic graft into a neoartery. *Nat Med* 18:1148–1153.

Wystrychowski, W., McAllister, T.N., Zagalski, K., Dusserre, N., Cierpka, L., and L'Heureux, N. 2014. First human use of an allogeneic tissue-engineered vascular graft for hemodialysis access. *J Vasc Surg* 60:1353–1357.

Zander, N.E., Orlicki, J.A., Rawlett, A.M., and Beebe, T.P. 2013. Electrospun polycaprolactone scaffolds with tailored porosity using two approaches for enhanced cellular infiltration. *J Mater Sci Mater Med* 24:179–187.

8 Biomolecular Modeling in Biomaterials

Sai J. Ganesan and Silvina Matysiak

CONTENTS

8.1 INTRODUCTION

Design of bioinspired fibrous materials has many applications both in the development of fundamental science, such as understanding protein folding, protein sequence to structure relationships, and self-assembly processes, and in engineering science, such as designing scaffolds for tissue engineering applications, nanofibers, and bioinspired devices among others. A majority of naturally existing scaffolds found throughout biology are protein-based, and these include actins, tubulins, and collagen, all of which have distinct properties that help in their function. Since synthesizing and working with such natural fibers are a challenge, over the last decade or more, there has been significant development in the *de novo* design of novel materials based on known protein sequence–structure relationships or design rules [1–4]. Binary design patterns of polar and nonpolar residues exist for the two major protein secondary structures: α-helices and β-strands [5,6]. Statistical analysis of folded structures has shown that nonpolar amino acids appear in the protein sequence, every three to four amino acids in α-helical structures, and every two amino acids in solvent-exposed β-sheets [7,8].

A balance between charge interactions, hydrogen bonding, and hydrophobicity is responsible for the formation of stable native protein folds [9–11]. Design of α-helical bundles or coiled coils with di-, tri-, and tetramers has led to the development of new fibrous materials [12]. These include fibers made of amphipathic α3 peptide [13], 28-residue *de novo* leucine-zipper peptides with sticky ends that promote longitudinal assembly [14], a trimeric coiled-coil motif TZ1H that forms helical fibrils as a consequence of a pH-induced conformational transition [15], and even a helix-turn-helix peptide of 18 residues that form self-associating helical fibers [16].

The most commonly recognized β-sheet fibrillar material is an amyloid-like structure, which is linked to a number of diseases from Alzheimer's to diabetes [17–19]. Multiple studies suggest that amyloid fibril structures are a generic structure accessible to all peptide chains, regardless of sequence specificity [20–24], and can be explained by the fact that backbone intermolecular bonds stabilize these structures and hence materials. This suggests that multiple bioinspired materials of varying physico-mechanical properties can be designed. Many amphipathic peptides like hydrophobins can self-assemble at air–water/oil–water or more generally hydrophobic–hydrophilic interfaces, changing their physical and mechanical properties like surface tension [25–27]. Thus, these peptides can be exploited in tissue engineering and surface design. Fibrillar systems of β-sheets can also form hydrogels [28], which have larger applications in drug delivery and tissue regeneration therapies [29], biosensors and microfluidic design [30], and scaffolds for tissue engineering [31].

Molecular simulations have been extremely useful in providing a molecular understanding of experimental observations, along with assisting new experimental peptide designs. Atomistic models of proteins provide most details but at a computational cost, due to which large time and length scales are inexplorable, whereas coarse-grained (CG) models help in understanding and differentiating between essential and approximate interactions. In addition, by reducing the level of resolution, CG models decrease the computational requirements compared to all-atom models and smoothen out the free-energy landscape facilitating sampling from one conformation to another. A wide range of CG models with varying resolutions are possible and have distinct uses. Minimalistic models where residues are represented by a few beads have been very valuable in advancing our understanding of the protein folding process [8,11,32,33]. But the main drawback in these models is that the parameters are tuned for a particular system and are not transferable to other systems. Therefore, they have to be recalibrated according to each system of interest. Some other models are intermediate resolution models, where the backbone is modeled with atomistic resolution and the side-chain (SC) coarse-grained. Upon inclusion of a hydrogen bonding interaction, these models are able to fold into helical structures without the addition of biases toward a native fold but fail toward stabilizing β-sheet structures [34,35]. The inclusion of explicit dipole–dipole interactions to these types of models has been shown to stabilize β-sheets [9,36,37]. However, all the models mentioned earlier renormalize the role of a solvent through effective short-range, inter-residue interactions and are hence not appropriate for studying the effect of the environment in protein folding and aggregation since an explicit solvent is needed. The recently developed water-explicit MARTINI CG force field [38] has been parameterized using water/oil and water/vacuum partitioning coefficients. By doing so, the model parameters are transferable and are not dependent on a particular system. This CG model captures many lipid membrane and transmembrane protein properties. However, it fails in capturing changes in the secondary structure and thus cannot be used to investigate protein folding.

In this chapter, we present a generic, minimalistic water-explicit polarizable protein model (WEPPROM) [39] that can be used to characterize the driving forces behind protein folding and aggregation without the addition of biasing potentials such as dihedral potentials or dependencies in the bending potential with secondary structure propensities. We use the binary design sequence patterns of helices (–HHPPHH–) and sheets (–HPHPHP–) [7] to fold into distinct secondary structures. We also use WEPPROM to understand the balance between hydrophobicity and backbone dipole interactions in ordered aggregation at the hydrophobic–hydrophilic interface and in water. This model can be used to design and test new *de novo* biomaterials for use in tissue engineering and other applications.

8.2 METHODS

We present a new water-explicit off-lattice CG protein model that can capture dipole interactions, analogous to the Drude oscillator approach [40,41]. We call our model WEPPROM [39] and it has roots in the MARTINI CG force-field [38]. Each polar CG bead that corresponds to elements that can hydrogen bond (H-bond) is modeled as a polarizable CG bead, as H-bonds are a type of dipole–dipole interactions. The spherical polarizable coarse-grained (pCG) bead consists of three interaction sites, the center bead, backbone bead (BB), and two dipole particles BBm and BBp, similar to the polarizable water model of Yesylevskyy et al. [42] and depicted in Figure 8.1.

Figure 8.1 Polarizable CG bead; the vdW radius of polarizable bead encloses the center of mass site BB and dummy particles BBm (negatively charged) and BBp (positively charged). The five tunable parameters (l, q, θ, k_l, k_θ) are depicted.

In WEPPROM, we mimic the directionality created by hydrogen bonds between the atoms CO and NH of a protein's backbone by modeling a backbone CG bead as a flexible dipole. An amino acid is modeled as two beads, a polarizable BB and a SC bead. The SC beads are broadly classified into hydrophobic (H), polar (P), and positively and negatively charged (Q+/Q–) as depicted in Figure 8.2. A single BB without a SC bead is defined as a neutral residue. The hydrophobic or polar characteristic of the residues is from pairwise potentials of SC beads. The CG protein model is combined with the recently developed polarizable CG water model [42]. The spherical backbone CG bead consists of three interaction sites as depicted in Figure 8.1. The dipole moment distribution of the protein BB is parameterized to match that of a peptide bond (3.5 D) [43].

The main CG site BB interacts with other CG beads through a pairwise Lennard-Jones (LJ) potential. Dipole particles BBm and BBp are harmonically bound to the central particle BB (distance [l], force constant [k_l]) and carry a positive and negative charge of equal magnitude q, respectively. These dipole particles interact with each other via electrostatic interactions. A harmonic angle potential (angle [θ] and angular force constant [k_θ]) is used to control the rotation of BBm and BBp particles. The location of the dipole particles is not fixed and hence the model is polarizable, that is, the orientation and moment of the backbone dipole bead are dependent on the electric field of the surrounding environment. A small repulsive core is added to the dipolar particles to avoid overpolarization. All 1–2 and 1–3 nonbonded interactions are excluded from both the main CG site and its embedded dipolar particles, as each main CG bead is covalently bonded by a harmonic bond potential to its nearest neighbors, and adjacent bonds are connected by a harmonic angle potential. Mass of each bead is 72 a.m.u., and it is distributed equally within the three sites, BB, BBm, and BBp.

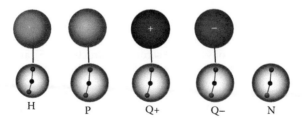

Figure 8.2 CG hydrophobic (H), polar (P), positively charged (Q+), negatively charged (Q–), and neutral (N) residues. Polarizable BB in black, enclosing dummy particles BBm and BBp, SC beads in cyan, green, blue, and red, respectively.

The force field in WEPPROM consists of both bonded parameters and nonbonded parameters. The bonded parameters include a harmonic angle and a bonded potential, between main CG sites as well as dummy particles inside the backbone CG bead. Dihedral potentials or biases are not used in WEPPROM. Nonbonded parameters include 12–6 LJ potential and Columbic potential:

$$U_{total} = U_{peptide-peptide} + U_{peptide-water} + U_{water-water}$$

$$U_{peptide-peptide} = U_{bonded} + U_{non\ bonded}$$

$$U_{peptide-water} = U_{non\ bonded}$$

$$U_{bonded} = U_{bonds} + U_{angles}$$

$$U_{non\ bonded} = U_{vdW} + U_{electrostatics}$$

$$U_{bonds} = \sum_{bonds} \frac{1}{2} k_l \left(l - l_o\right)^2$$

$$U_{angles} = \sum_{angles} \frac{1}{2} k_\theta \left(\theta - \theta_o\right)^2$$

$$U_{vdW} = \sum_{pairs(i,j)} 4\epsilon_{ij} \left[\left(\frac{\sigma_{ij}}{r_{ij}}\right)^{12} - \left(\frac{\sigma_{ij}}{r_{ij}}\right)^6\right]$$

$$U_{electrostatics} = \sum_{pairs(i,j)} \frac{q_i q_j}{4\pi\epsilon_0\epsilon_r r_{ij}}$$

where
U denotes potential energy
r is distance between beads
$k_{l/\theta}$ is bond or bead angle strength
l_o/θ_o is equilibrium bond length or bead angle
l/θ is bond length or bead angle
σ_{ij} is the closest distance of approach between two beads or effective size
ϵ_{ij} is the strength of interaction between two beads
q is the charge on the bead

A bond is defined between two adjacent BBs or between a BB and its SC bead. An angle is defined between three or two BBs and a SC bead. The *pairs(i, j)* for peptide–peptide interactions are different from the ones for peptide–water interactions. For all model parameters, parameterization details, and simulation parameters, see References 39 and 44. For details on $U_{water-water}$, see Reference 42.

8.3 APPLICATIONS

8.3.1 Folding of Supersecondary Structures

The formation of a protein backbone's hydrogen bonds leads to a distinct orientation of dipoles in various types of secondary structures [45], and protein engineering studies have identified patterns of hydrophobic and polar residues to design helix and sheet bundles [7,8]. Hence, to test the ability of WEPPROM to capture distinct protein secondary structures accurately, we used prominent peptide design sequence patterns for helices (–HHPPHH–) and sheets (–HPHPHP–) [7] to design helix and sheet bundles with the inclusion of turning regions.

A 12-peptide repeating sequence pattern of –HHPP– folds into a helix without any bias. Since hydrophilic loops are common in proteins, we use polar turning residues for both helix and sheet bundles [46–48]. Folding of a 33-residue helix

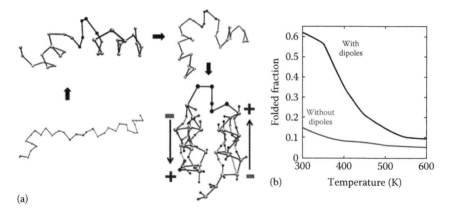

(a)

(b)

Figure 8.3 (a) Stages in helix bundle formation, starting from a coil-like conformation. Only BBs are shown (green for polar and cyan for hydrophobic residues) and (b) temperature-dependent folding curves for the helix bundle with and without dipolar particles. (Reprinted with permission from Ganesan, S.J. and Matysiak, S., Role of backbone dipole interactions in the formation of secondary and supersecondary structures of proteins, *J. Chem. Theory Comput.*, 10(6), 2569–2576. Copyright 2014 American Chemical Society.)

bundle sequence $(HHPP)_3HH-(T)_5-(PPHH)_3PP$ from a completely extended conformation is shown in Figure 8.3a. The chain folds into two distinct helices on either ends in a hierarchical manner, followed by the collapse of the chain to form the α-helix bundle. Folding of the helices on either ends is induced by backbone dipole interactions and their subsequent orientations, which is a local effect. Therefore, the secondary structure is sufficiently stable in the absence of tertiary interactions, and the formation of secondary and supersecondary structures is uncoupled. Hierarchical folding has been observed for several proteins, such as the engrailed homeodomain [49,50]. However, once backbone dipole particles are removed, the peptide does not fold into a helix bundle and remains collapsed, with no helicity. Without dipole particles, the folded fraction computed as a measure of helical dihedrals remains at 0.1 for all temperatures between 300 and 600 K (red curve in Figure 8.3b). When backbone dipoles are present, the folding process becomes cooperative, achieving a folded fraction of 0.6 at 300 K, as evident from the sigmoidal shape of the blue folding curve in Figure 8.3b. Folding of helix bundles is driven by both dipolar interactions and pairwise LJ interactions, that is, a combination of the sequence patterning and electrostatic interactions. However, the role of electrostatics is more significant in the systems we explored. The macrodipoles of each of the two helices, which is computed by adding the microdipoles of each helix (denoted in Figure 8.3a), signal an antiparallel dipole orientation that might provide additional favorable electrostatic interaction energy for the formation of helix bundles. The known helix bundles of protein structures in the literature exhibit similar antiparallel orientation [51].

We then tested the ability of the model in predicting β-strand bundles, using two polar turn residues [52–54] and the sheet pattern –HPHP–. The 19-residue sequence pattern folds without any biases into a three-strand antiparallel β-sheet consisting of two hairpins. This structure is reminiscent to the $\beta3s$ protein [55,56] and WW domain [57], and this sequence pattern is common in solvent-exposed β-sheets [7]. The folding of the three-stranded β-sheet also occurs in the

177

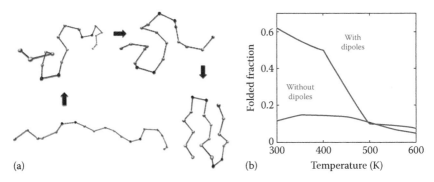

Figure 8.4 (a) Stages in sheet bundle formation, starting from a coil-like conformation. Only BBs are shown (green for polar and cyan for hydrophobic residues) and (b) temperature-dependent folding curves for sheet bundles with and without dipolar particles. (Reprinted with permission from Ganesan, S.J. and Matysiak, S., Role of backbone dipole interactions in the formation of secondary and supersecondary structures of proteins, *J. Chem. Theory Comput.*, 10(6), 2569–2576. Copyright 2014 American Chemical Society.)

hierarchical order of hairpin formation (Figure 8.4a). One hairpin is formed first, with the second hairpin being formed at the last stage of folding, in agreement with all-atom simulations of β3s [55] and WW domain [57]. The matching hydrophobic and polar residues in the folded structure are also worth noticing. Shorter –HP– sequences with one turn and four stranded β-sheets with three turning regions also fold into sheet bundles. When dipole particles are removed, sheet bundles are not formed and no sheet content is observed. The folding curve for peptide with and without dipole particles is depicted in Figure 8.4b, with the peptide with dipolar particles exhibiting a cooperative folding–unfolding transition. On comparing electrostatic stability between helices and sheets by evaluating dipole energies [39], helical conformations were found to be more stabilized by the backbone dipoles than sheets. Furthermore, helices were better aligned with the electric field than sheets.

8.3.2 Ordered Aggregation of Octapeptides

Protein aggregation is a hallmark of multiple diseases such as Alzheimer's, Huntington's, and prion diseases, among others [17–19], and designed β-sheet fibrils have been extensively used to mimick biological materials for drug delivery and tissue engineering applications [58–61]. Multiple studies suggest that the ability of proteins and peptides to form amyloid fibril structures is not limited to disease conditions, but a generic property of all polypeptides [20–22]. We have used WEPPROM to understand the balance between hydrophobicity and dipole interactions in peptide aggregation, both in water and at the water–hexadecane interface. Hexadecane or octane slabs are known to be a good membrane mimetic environment [62]. We have chosen $(GV)_4$ octapeptide as a model system for this study as experimental and simulation studies have shown $(GV)_n$ peptides to be good models for studying amyloid formation, and these peptides have a propensity for forming β-sheet-containing fibrils both in water and at octane–water interfaces [63,64].

Ordered aggregation both in water and at the hydrophobic–water interface was observed at high peptide concentrations with dipolar interactions playing a more significant role in interfaces. The aggregates formed on the

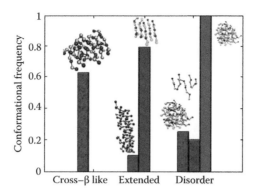

Figure 8.5 Conformational frequency of a high peptide concentrated system in water (red), at the interface (blue), and the control system without dipolar particles in a water environment (green). Representative conformations are shown as insets, and BBs are represented in cyan (hydrophobic residue) and purple (neutral residue). Hydrophobic SC beads are shown in green. (Reproduced from Ganesan, S.J. and Matysiak, S., Interplay between the hydrophobic effect and dipole interactions in peptide aggregation at interfaces, *Phys. Chem. Chem. Phys.*, 18, 2449–2458, 2016. With permission from the PCCP Owner Societies.)

interface are more ordered than aggregates in water. Figure 8.5 compares the conformational frequency between aggregates in water (red bars), in the presence of a hydrophobic interface (blue bars), and systems without dipolar particles (green bar). The presence of a hydrophobic surface influences aggregation morphology by shifting the conformational equilibrium. In water, two distinct types of conformations were observed: (1) a cross-β-like conformation with a frequency of 0.6, and (2) an extended fibril-like conformation, with a frequency of 0.1. Studies on $A\beta_{1-40}$ show that within 6–10 monomer chains, the cross-β-like order is seen [65]. The "fibril-like" conformation is more water exposed than the "cross-β-like" conformation. The "cross-β-like" conformation observed is similar to the MVGGVV microstructure crystallized by Eisenberg's group [66]. With the presence of a hydrophobic surface, there is a conformational shift to "fibril-like" structures, with the hydrophobic beads partitioning into the hexadecane phase. The frequency of ordered, extended conformations in the presence of an interface is 0.8. Another important feature is the decrease in disordered conformations by about 10% in the biphasic system compared to bulk water. Partitioning of valine or hydrophobic residues into hexadecane restricts or constrains the peptides to two dimensions. Insets in Figure 8.5 show representative conformations.

At 300 K, we see unidirectional progression of aggregation or a downhill mechanism, and an increased rate of ordered aggregation in the biphasic system compared to water is observed. The aggregation in water followed a two-step condensation-ordering mechanism, where randomly distributed peptides in water initially collapse into disordered aggregates with the formation of a hydrophobic core [44]. A comparison of kinetic properties of aggregation in water and the more probable aggregation pathway in biphasic systems is depicted in Figure 8.6b. In water, the monomers (M) (solid pink curve) aggregate to form an intermediate disordered state (I) (solid cyan curve), which after 25 ns, reorganizes to form ordered sheets (O) (solid black curve). Representative conformations are shown in Figure 8.6c. However, when the water–hexadecane surface is present, the monomers form sheets in less than 10 ns (dashed black

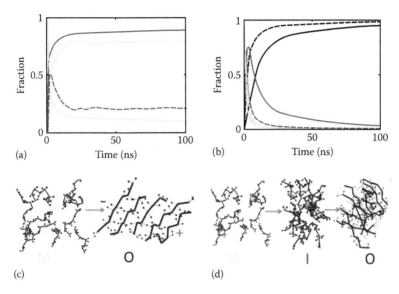

(a) Time (ns)

(b) Time (ns)

(c)

(d)

Figure 8.6 Time evolution of the fraction of different species for peptide aggregation (a) without dipolar particles and (b) with dipolar particles. Dashed lines represent system with interface and solid lines, system in water. Black lines represent ordered aggregates [O]; pink, disordered fraction [I]; and monomer fraction in cyan [M]. (c) Representative pathway for ordered aggregate formation at the hydrophobic–water interface, monomers [M] form ordered aggregates [O] with shortly lived intermediates. (d) Representative pathway for ordered aggregate formation in explicit water, monomers [M] form intermediate aggregates [I] that rearrange to form ordered conformation [O]. The peptide backbone is shown in black. (Reproduced from Ganesan, S.J. and Matysiak, S., Interplay between the hydrophobic effect and dipole interactions in peptide aggregation at interfaces, *Phys. Chem. Chem. Phys.*, 18, 2449–2458, 2016. With permission from the PCCP Owner Societies.)

curve in Figure 8.6b). The nature of the ordered fraction curves (dashed black curve in Figure 8.6b) is more exponential in the biphasic system than in water (solid black curve in Figure 8.6b), implying a more cooperative process. Representative conformations of ordered aggregation at interface and in water are shown in Figure 8.6c and d, respectively. Thus, the presence of the water–hydrophobic interface aids or accelerates the rate of sheet formation, relative to water, as has been observed experimentally [67]. The interfacial restriction of peptides increases the probability for collision as the local concentration increases, which can in turn lead to an increase in peptide–peptide hydrogen bonds and hence aggregation. When dipole particles are removed from the system of peptides in water, the monomers (solid cyan curve in Figure 8.6a) interact to form disordered aggregates (solid pink curve in Figure 8.6a). Since there are no dipolar interactions, there is no order in the aggregate. The ordered conformational fraction remains at 0 (solid black curve in Figure 8.6a). In the presence of an interface, the peptides without dipolar interactions get adsorbed to the surface, due to partitioning of hydrophobic side chains. However, due to the lack of interpeptide dipole interactions, the peptides do not aggregate on the interface and are free to diffuse individually. Since the hydrophobic side chain can partition to hexadecane, the entropy is driving the peptides to not aggregate. Based on this, we conclude that at hydrophobic interfaces, the major driving force for peptide aggregation is dipolar interactions, which in our model mimics hydrogen bonding. Figure 8.6a (dashed lines) further elucidates this point. The monomer fraction (cyan dashed curve) decreases as the peptides aggregate and reach the interface; however, there is an increase soon after 5 ns. This is due to the fact that the monomers without dipolar particles, on reaching the interface, are free to diffuse. Since the hydrophobic side chains remain buried in hexadecane, there is no other driving force to keep the peptides aggregated. However, when dipole particles are present, ordered aggregates are formed within the first 10 ns.

8.4 CONCLUSIONS

Significant progress on CG models has been made in the last decade. In this chapter, we have highlighted a new model we have developed and its potential applications in the study of molecular interactions in biological systems. The WEPPROM model currently accounts for only five different kinds of residues, and our lab is working on establishing a CG scheme and parameter set for all 20 amino acids. We have recently extended the polarizable model to lipids to account for anionic and zwitterionic lipids by adding dipole particles inside CG beads that can form H-bonds. Our lab has recently successfully integrated WEPPROM with a membrane model to study peptide–lipid bilayer interactions [68]. The results obtained using our model are in excellent agreement with various other studies, but also provide new physicochemical insights into the nature of molecular interactions. In this chapter, we have only glanced at two specific applications of WEPPROM; however, since our model is sensitive to environmental changes and external fields, it has tremendous scope in being adapted to other biomaterials, both peptide-based and otherwise. WEPPROM can also be extended to either model natural materials or engineer and design new fibrous biomaterials. Since these materials are more biocompatible, they could be used in engineering sensors and devices for biomedical applications.

REFERENCES

1. D. N. Woolfson, The design of coiled-coil structures and assemblies, *Adv. Protein Chem.*, 70, 79–112, 2005.

2. S. A. Potekhin, T. Melnik, V. Popov, N. Lanina, A. A. Vazina, P. Rigler, A. Verdini, G. Corradin, and A. Kajava, De novo design of fibrils made of short α-helical coiled coil peptides, *Chem. Biol. (Oxford, UK)*, 8(11), 1025–1032, 2001.

3. A. Aggeli, I. Nyrkova, M. Bell, R. Harding, L. Carrick, T. McLeish, A. Semenov, and N. Boden, Hierarchical self-assembly of chiral rod-like molecules as a model for peptide β-sheet tapes, ribbons, fibrils, and fibers, *Proc. Natl. Acad. Sci. USA*, 98(21), 11857–11862, 2001.

4. D. J. Pochan, J. P. Schneider, J. Kretsinger, B. Ozbas, K. Rajagopal, and L. Haines, Thermally reversible hydrogels via intramolecular folding and consequent self-assembly of a de novo designed peptide, *J. Am. Chem. Soc.*, 125(39), 11802–11803, 2003.

5. S. Kamtekar, J. M. Schiffer, H. Xiong, J. M. Babik, and M. H. Hecht, Protein design by binary patterning of polar and nonpolar amino acids, *Science*, 262(5140), 1680–1685, 1993.

6. Y. Wei, T. Liu, S. L. Sazinsky, D. A. Moffet, I. Pelczer, and M. H. Hecht, Stably folded de novo proteins from a designed combinatorial library, *Protein Sci.*, 12(1), 92–102, 2003.

7. M. W. West and M. E. Hecht, Binary patterning of polar and nonpolar amino acids in the sequences and structures of native proteins, *Protein Sci.*, 4, 2032–2039, 1995.

8. G. Bellesia, A. I. Jewett, and J.-E. Shea, Sequence periodicity and secondary structure propensity in model proteins, *Protein Sci.*, 19, 141–154, 2009.

9. T. Bereau and M. Deserno, Generic coarse-grained model for protein folding and aggregation, *J. Chem. Phys.*, 130, 235106, 2009.

10. S. Matysiak and P. Das, Effect of sequence and solvation on the temperature–pressure conformational landscape of proteinlike heteropolymers, *Phys. Rev. Lett.*, 111, 058103, 2013.

11. J. N. Onuchic, Z. Luthey-Schulten, and P. G. Wolynes, Theory of protein folding: The energy landscape perspective, *Annu. Rev. Phys. Chem.*, 48, 545–600, 1997.

12. W. D. Kohn and R. S. Hodges, De novo design of α-helical coiled coils and bundles: Models for the development of protein-design principles, *Trends Biotechnol.*, 16(9), 379–389, 1998.

13. S. Kojima, Y. Kuriki, T. Yoshida, K. Yazaki, and K.-I. Miura, Fibril formation by an amphipathic α-helix-forming polypeptide produced by gene engineering, *Proc. Jpn. Acad. Ser. B*, 73(1), 7–11, 1997.

14. M. J. Pandya, G. M. Spooner, M. Sunde, J. R. Thorpe, A. Rodger, and D. N. Woolfson, Sticky-end assembly of a designed peptide fiber provides insight into protein fibrillogenesis, *Biochemistry*, 39(30), 8728–8734, 2000.

15. Y. Zimenkov, S. N. Dublin, R. Ni, R. S. Tu, V. Breedveld, R. P. Apkarian, and V. P. Conticello, Rational design of a reversible pH-responsive switch for peptide self-assembly, *J. Am. Chem. Soc.*, 128(21), 6770–6771, 2006.

16. K. L. Lazar, H. Miller-Auer, G. S. Getz, J. P. Orgel, and S. C. Meredith, Helixturn-helix peptides that form α-helical fibrils: Turn sequences drive fibril structure, *Biochemistry*, 44(38), 12681–12689, 2005.

17. D. Burdick, B. Soreghan, M. Kwon, J. Kosmoski, M. Knauer, A. Henschen, J. Yates, C. Cotman, and C. Glabe, Assembly and aggregation properties of synthetic Alzheimer's A4/β amyloid peptide analogs, *J. Biol. Chem.*, 267(1), 546–554, 1992.

18. E. Scherzinger, A. Sittler, K. Schweiger, V. Heiser, R. Lurz, R. Hasenbank, G. P. Bates, H. Lehrach, and E. E. Wanker, Self-assembly of polyglutamine-containing huntingtin fragments into amyloid-like fibrils: Implications for huntington's disease pathology, *Proc. Natl. Acad. Sci. USA*, 96(8), 4604–4609, 1999.

19. F. Chiti and C. M. Dobson, Protein misfolding, functional amyloid, and human disease, *Annu. Rev. Biochem.*, 75, 333–366, 2006.

20. D. Thirumalai, D. Klimov, and R. Dima, Emerging ideas on the molecular basis of protein and peptide aggregation, *Curr. Opin. Struct. Biol.*, 13(2), 146–159, 2003.

21. M. Stefani and C. M. Dobson, Protein aggregation and aggregate toxicity: New insights into protein folding, misfolding diseases and biological evolution, *J. Mol. Med.*, 81(11), 678–699, 2003.

22. C. M. Dobson, Principles of protein folding, misfolding and aggregation, *Semin. Cell Dev. Biol.*, 15(1), 3–16, 2004.

23. T. Koga, K. Taguchi, Y. Kobuke, T. Kinoshita, and M. Higuchi, Structural regulation of a peptide-conjugated graft copolymer: A simple model for amyloid formation, *Chem. Eur. J.*, 9(5), 1146–1156, 2003.

24. C. E. MacPhee and C. M. Dobson, Formation of mixed fibrils demonstrates the generic nature and potential utility of amyloid nanostructures, *J. Am. Chem. Soc.*, 122(51), 12707–12713, 2000.

25. J. P. Mackay, J. M. Matthews, R. D. Winefield, L. G. Mackay, R. G. Haverkamp, and M. D. Templeton, The hydrophobin eas is largely unstructured in solution and functions by forming amyloid-like structures, *Structure*, 9(2), 83–91, 2001.

26. H. Wösten, F. Schuren, and J. Wessels, Interfacial self-assembly of a hydrophobin into an amphipathic protein membrane mediates fungal attachment to hydrophobic surfaces, *EMBO J.*, 13(24), 5848, 1994.

27. H. A. Wosten, O. M. De Vries, and J. G. Wessels, Interfacial self-assembly of a fungal hydrophobin into a hydrophobic rodlet layer, *Plant Cell*, 5(11), 1567–1574, 1993.

28. C. Yan, A. Altunbas, T. Yucel, R. P. Nagarkar, J. P. Schneider, and D. J. Pochan, Injectable solid hydrogel: Mechanism of shear-thinning and immediate recovery of injectable β-hairpin peptide hydrogels, *Soft Matter*, 6(20), 5143–5156, 2010.

29. N. Peppas, P. Bures, W. Leobandung, and H. Ichikawa, Hydrogels in pharmaceutical formulations, *Eur. J. Pharm. Biopharm.*, 50(1), 27–46, 2000.

30. A. K. Yetisen, I. Naydenova, F. da Cruz Vasconcellos, J. Blyth, and C. R. Lowe, Holographic sensors: Three-dimensional analyte-sensitive nanostructures and their applications, *Chem. Rev.*, 114(20), 10654–10696, 2014.

31. A. Mellati, S. Dai, J. Bi, B. Jin, and H. Zhang, A biodegradable thermosensitive hydrogel with tuneable properties for mimicking three-dimensional microenvironments of stem cells, *RSC Adv.*, 4(109), 63951–63961, 2014.

32. S. Matysiak and C. Clementi, Mapping folding energy landscapes with theory and experiment, *Arch. Biochem. Biophys.*, 468, 29–33, 2008.

33. T. Head-Gordon and S. Brown, Minimalist models for protein folding and design, *Curr. Opin. Struct. Biol.*, 13, 160–167, 2003.

34. S. Takada, Z. Luthey-Schulten, and P. G. Wolynes, Folding dynamics with nonadditive forces: A simulation study of a designed helical protein and a random heteropolymer, *J. Chem. Phys.*, 110, 11616, 1999.

35. A. Irback, F. Sjunnesson, and S. Wallin, Three-helix-bundle protein in a Ramachandran model, *Proc. Natl. Acad. Sci. USA*, 97, 13614–13618, 2000.

36. Y. Mu and Y. Q. Gao, Effect of hydrophobic and dipole-dipole interactions on the conformational transitions of a model polypeptide, *J. Chem. Phys.*, 127, 105102, 2007.

37. N.-Y. Chen, Z.-Y. Su, and C.-Y. Mou, Effective potentials for protein folding, *Phys. Rev. Lett.*, 96, 078103, 2006.

38. L. Monticelli, S. K. Kandasamy, X. Periole, R. G. Larson, D. P. Tieleman, and S.-J. Marrink, The MARTINI coarse grained forcefield: Extension to proteins, *J. Chem. Theory Comput.*, 4, 819–834, 2008.

39. S. J. Ganesan and S. Matysiak, Role of backbone dipole interactions in the formation of secondary and supersecondary structures of proteins, *J. Chem. Theory Comput.*, 10(6), 2569–2576, 2014.

40. G. Lamoureux and B. Roux, Modeling induced polarization with classical drude oscillators: Theory and molecular dynamics simulation algorithm, *J. Chem. Phys.*, 119(6), 3025–3039, 2003.

41. S. Patel, A. D. Mackerell, and C. L. Brooks, CHARMM fluctuating charge force field for proteins: II protein/solvent properties from molecular dynamics simulations using a nonadditive electrostatic model, *J. Comput. Chem.*, 25(12), 1504–1514, 2004.

42. S. O. Yesylevskyy, L. V. Schäfer, D. Sengupta, and S. J. Marrink, Polarizable water model for the coarse-grained martini force field, *PLoS Comput. Biol.*, 6(6), e1000810, 2010.

43. T. E. Creighton, *Proteins: Structures and Molecular Properties*, 2nd edn. W.H. Freeman, New York, 1992.

44. S. J. Ganesan and S. Matysiak, Interplay between the hydrophobic effect and dipole interactions in peptide aggregation at interfaces, *Phys. Chem. Chem. Phys.*, 18, 2449–2458, 2016.

45. W. G. J. Hol, L. M. Halie, and C. Sander, Dipoles of the α-helix and β-sheet: Their role in protein folding, *Nature*, 294, 532–536, 1981.

46. C. Wilmot and J. Thornton, Analysis and prediction of the different types of β-turn in proteins, *J. Mol. Biol.*, 203(1), 221–232, 1988.

47. A. V. Efimov, Patterns of loop regions in proteins, *Curr. Opin. Struct. Biol.*, 3(3), 379–384, 1993.

48. M. W. West, W. Wang, J. Patterson, J. D. Mancias, J. R. Beasley, and M. H. Hecht, De novo amyloid proteins from designed combinatorial libraries, *Proc. Natl. Acad. Sci. USA*, 96(20), 11211–11216, 1999.

49. N. D. Clarke, C. R. Kissinger, J. Desjarlais, G. L. Gilliland, and C. O. Pabo, Structural studies of the engrailed homeodomain, *Protein Sci.*, 3(10), 1779–1787, 1994.

50. V. Daggett and A. R. Fersht, Is there a unifying mechanism for protein folding? *Trends Biochem. Sci.*, 28(1), 18–25, 2003.

51. R. P. Sheridan, R. M. Levy, and F. R. Salemme, α-Helix dipole model and electrostatic stabilization of 4-α-helical proteins, *Proc. Natl. Acad. Sci. USA*, 79, 4545–4549, 1982.

52. H. E. Stanger and S. H. Gellman, Rules for antiparallel, β-sheet design: D-pro-gly is superior to L-asn-gly for, β-hairpin nucleation, *J. Am. Chem. Soc.*, 120, 4236–4237, 1998.

53. E. D. Alba, J. Santoro, M. Rico, and M. A. Jiménez, De novo design of a monomeric three-stranded antiparallel-sheet, *Protein Sci.*, 8(4), 854–865, 1999.

54. S. H. Gellman, Minimal model systems for β-sheet secondary structure in proteins, *Curr. Opin. Struct. Biol.*, 2(6), 717–725, 1998.

55. S. V. Krivov, S. Muff, A. Caflisch, and M. Karplus, One-dimensional barrier-preserving free-energy projections of a beta-sheet miniprotein: New insights into the folding process, *J. Phys. Chem. B*, 112(29), 8701–8714, 2008.

56. J. M. Carr and D. J. Wales, Folding pathways and rates for the three-stranded betasheet peptide beta3s using discrete path sampling, *J. Phys. Chem. B*, 112(29), 8760–8769, 2008.

57. S. Beccara, T. Skrbić, R. Covino, and P. Faccioli, Dominant folding pathways of a WW domain, *Proc. Natl. Acad. Sci. USA*, 109(7), 2330–2335, 2012.

58. N. L. Goeden-Wood, J. D. Keasling, and S. J. Muller, Self-assembly of a designed protein polymer into β-sheet fibrils and responsive gels, *Macromolecules*, 36(8), 2932–2938, 2003.

59. W. A. Petka, J. L. Harden, K. P. McGrath, D. Wirtz, and D. A. Tirrell, Reversible hydrogels from self-assembling artificial proteins, *Science*, 281(5375), 389–392, 1998.

60. A. P. Nowak, V. Breedveld, L. Pakstis, B. Ozbas, D. J. Pine, D. Pochan, and T. J. Deming, Rapidly recovering hydrogel scaffolds from self-assembling diblock copolypeptide amphiphiles, *Nature*, 417(6887), 424–428, 2002.

61. J. P. Schneider, D. J. Pochan, B. Ozbas, K. Rajagopal, L. Pakstis, and J. Kretsinger, Responsive hydrogels from the intramolecular folding and self-assembly of a designed peptide, *J. Am. Chem. Soc.*, 124(50), 15030–15037, 2002.

62. R. M. Venable, Y. Zhang, B. J. Hardy, and R. W. Pastor, Molecular dynamics simulations of a lipid bilayer and of hexadecane: An investigation of membrane fluidity, *Science*, 262(5131), 223–226, 1993.

63. M. Seo, S. Rauscher, R. Pomès, and D. P. Tieleman, Improving internal peptide dynamics in the coarse-grained martini model: Toward large-scale simulations of amyloid and elastin-like peptides, *J. Chem. Theory Comput.*, 8(5), 1774–1785, 2012.

64. A. Nikolic, S. Baud, S. Rauscher, and R. Pomès, Molecular mechanism of β-sheet self-organization at water–hydrophobic interfaces, *Proteins*, 79(1), 1–22, 2011.

65. S. Brown, N. J. Fawzi, and T. Head-Gordon, Coarse-grained sequences for protein folding and design, *Proc. Natl. Acad. Sci. USA*, 100(19), 10712–10717, 2003.

66. M. R. Sawaya, S. Sambashivan, R. Nelson, M. I. Ivanova, S. A. Sievers, M. I. Apostol, M. J. Thompson et al., Atomic structures of amyloid cross-β spines reveal varied steric zippers, *Nature*, 447(7143), 453–457, 2007.

67. G. W. Vandermeulen, K. T. Kim, Z. Wang, and I. Manners, Metallopolymerpeptide conjugates: Synthesis and self-assembly of polyferrocenylsilane graft and block copolymers containing a beta-sheet forming Gly-Ala-Gly-Ala tetrapeptide segment, *Biomacromolecules*, 7(4), 1005–1010, 2006.

68. S. J. Ganesan, H. Xu, and S. Matysiak, Effect of lipid head group interactions on membrane properties and membrane-induced cationic β-hairpin folding, *Phys. Chem. Chem. Phys.*, 18, 17836–17850, 2016.

9 Finite Element Analysis in Biomaterials

Clark A. Meyer

CONTENTS

9.1 INTRODUCTION

As noted in other chapters, biomaterials are the materials that are used in the body, including implanted substances from external sources. Tissue failures and medical device failures from structures constructed of these materials are often costly and can be traced to design flaws. Finite element analyses can supplement or replace traditional bench testing to provide greater insight into performance questions, including the likelihood of medical device implant success or failure due to structural problems.

Finite element analysis (FEA) is a computational technique for calculating equilibrium. FEA is often applied to the calculation of stresses and strains of objects in response to structural loading at (quasi-) static equilibrium, but potentially under dynamic loading as well. FEA was primarily developed for structural analyses that demanded efficiency of material use including assessing automobiles and airplanes, but FEA has been used more broadly as a foundational approach behind multi-physics analyses that incorporate more than just mechanical structural loads, but also thermal, electromagnetic, and/or fluid loads. As computing power has increased and become cheaper, the usage of FEA has spread; FEA is now widely used during the development of medical devices.

This chapter will present a simple understanding of FEA and how it works, a basic appreciation for how biomaterials are used in FEA, and areas of recent application. Citations to a few sources are provided throughout, particularly for particular biomedical applications, but for details on the overall method, interested readers are referred to full textbooks on the topic (Kwon and Bang 2000, Bathe 2006, MacDonald 2011, Reddy and Reddy 2015).

9.2 COMPONENTS OF FINITE ELEMENT ANALYSIS

Finite element analysis is a computational methodology for approximating solutions to equations over possibly complex shapes by representing the complex shape as a collection of simpler shapes, where the solution can be approximated, and then ensuring continuity of the solution between the simpler shapes. Put less precisely, FEA is using a computer to divide up the object of interest into regular chunks, add necessary information about the materials and conditions, and solve for equilibrium displacements and stresses in the object in response to being loaded.

Conducting a typical FEA requires

1. Clear identification of the goal of the analysis

2. Recognition of the equations to be solved (e.g., the general equations of structural mechanics)

3. Geometry (i.e., "shapes") over which the equations will be solved

4. Material properties of the materials the shapes consist of

5. Meshing to subdivide the shapes into simple shapes (called "elements") over which the solution can be approximated

6. Load steps that describe the loads (e.g., forces)

7. Boundary conditions (e.g., fixations or pins) being applied to the geometry

8. A solver

9. Post-processing to extract the decisive information

The goal of an analysis, or of a series of analyses, influences many of the decisions that are made in the design and setup of the analysis. Goals for FEA in biomaterials are typically related to predictions of failure (often from overstressing or from fatigue). There have been applications where the goal is instead either to determine information that feeds into other analyses, which require information on the stress or strain of the material (Huiskes et al. 2000) as they can impact growth and remodeling of tissue, or to determine the relationship between stress and strain in a multi-scale modeling effort (Hayenga et al. 2014) where an FEA model is used as a sub-model within another model or FEA, or as a part of an optimization design step where the stresses vary with design parameters.

In light of the overall goal of analyses, simplifications can be made to the geometry, material properties, loads, and boundary conditions to make the analysis or analyses easier to interpret, faster to run, and more useful. When conducting FEA for use in FDA submissions, draft guidance explaining required contents of the reporting is available (FDA 2014) with final, official guidance from the FDA expected in late 2016. Also worth considering at the outset of an analysis is how verification, validation, and uncertainty quantification will be done to ensure the modeling effort is fully reliable (Pathmanathan and Gray 2013).

Equation 9.1 is typically being solved in a FEA based on structural mechanics,

$$ku = F \tag{9.1}$$

where
  F is force
  k is stiffness
  u is displacement

In a linear system, each of these terms is separate. However, in many problems of interest to those working with biomaterials, the system is not linear because there are interdependencies between terms. A common dependency is the change of observed stiffness of the biomaterial due to a change in displacement. This change in stiffness can be either from the geometric effects of substantial rotations or due to the nonlinear material properties of components within the biomaterial. A linear system directly relating force, stiffness, and displacement can be solved quickly and simply but can yield inaccurate results. Unit system consistency is crucial to the accuracy of a finite element model, and many programs rely on the user to be appropriately self-consistent (e.g., all standard SI) such that units and thus unit conversions are never directly specified within the program.

Once recognition of the type of analysis and of requisite equations has been made, the next step is to generate the geometry of the object that is to be the subject of analysis. As an example, for the case of peripheral stent design, this geometry might consist of merely a portion of the stent itself or perhaps the stent and involved parts (i.e., the stent, its delivery system, and a virtual artery). The scope of the geometry to be included in an analysis depends on the aim of the modeling effort as well as on possible symmetries to reduce computational requirements and usage. If the goal is simply to assess fatigue response to known circumferential loading, simpler models can suffice. "Defeaturing" is the removal of excessive detail from a model and allows for faster, simpler models that still achieve the necessary accuracy—recognition of what is excessive depends on engineering judgment and potentially upon testing (real and/or virtual) as well. When using a developed FEA software package (common ones used with biomaterials include Abaqus, Ansys, FEbio, and COMSOL), geometry can be imported from computer-aided design (CAD) files, imported from processed medical images (using tools such as Mimics/3matic or ScanIP/scanFE), or created within the finite element program's built-in CAD tools.

Once the geometry of the component(s) to be analyzed is in place within the software, meshing can be proceeded. Meshing is the process whereby the complex (or at least potentially complex) geometry is divided into pieces of relatively regular shapes—see Figures 9.1 and 9.2 for an example geometry and mesh. These regular shapes are the elements of the finite element method and, in 3D models, commonly consist of tetrahedrals (four-sided pyramids), wedges, and hexahedrals (six-sided blocks). Other shapes are possible and sometimes used in special circumstances. For first-order elements, nodes are located at the corners of the elements. First order indicates that the degree of interpolation of the sides is of the first order (straight). The nodes of elements are numbered in two ways, by a local system, which defines the orientation of

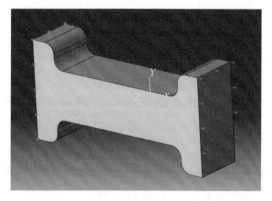

Figure 9.1 Capture of geometry and boundary conditions for a simple FEA of a dog-bone-shaped sample with uniaxial displacement. Small red cones at the left face indicate fixation in x, y, and z directions. Red arrows on the right face indicate the direction and the degree of displacement to be applied.

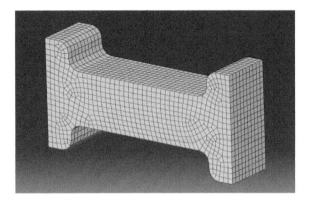

Figure 9.2 Example of hexahedral mesh applied to dog-bone geometry.

the element and its faces, and a global system, which provides unique identifiers to all nodes within the model. Elements are also numbered, so a bookkeeping system (often hidden from the user) of each element can be used to track the mesh. This bookkeeping system contains information regarding the nodal positions, comprised of global node number and position in three space dimensions (x, y, and z coordinates) in conjunction with connectivity, a listing consisting of global element number and associated nodes (each identified with their global node number), listed in the order (such that orientation information is captured) of each element. The equations will be evaluated over a global mesh, and the quality of the mesh can greatly influence the quality of the solution. Quality of a mesh is characterized by the element shape and density. Shapes that are closer to ideals (e.g., equilateral tetrahedral elements, cubic hexahedral elements) typically perform better because they avoid mathematical difficulties. The density of a mesh is related to the interpolation of the solution, a greater density (i.e., more, smaller elements) reduces the degree of interpolation and typically provides a more accurate answer. However, increasing mesh density increases computational cost and in very fine meshes, numeric precision errors can prorogate and reduce the accuracy. The nodes are where continuity of the solution is enforced between elements. Since higher-order elements have additional nodes on their sides, they represent curved surfaces more easily (i.e., the same accuracy with fewer elements). The mesh seeding operation controls the density of the mesh by prescribing density (nodal spacing) rules on edges or globally.

Material properties within the context of structural FEA are how the relationships between stresses and strains are defined for the objects being analyzed. Specifically, specification of the material properties goes to the definition of the stiffness variable (K). Calculation of the value of K can be decomposed into the part from the material and the part from the geometry itself. Technically speaking, for a 3D analysis, the fourth-order elasticity tensor relates each component of the second-order strain tensor to that of the second-order stress tensor. However, in many cases, this 81-component fourth-order tensor can be simplified greatly due to symmetries and/or homogeneity. Most FEA programs have the ability to use a wide range of materials by the user simply supplying the constants instead of the components of the elasticity tensor itself. The user might merely type in elastic constants such as Young's Modulus and Poisson's ratio for a material being represented with linear elastic formulation.

It is important to keep in mind the goal of the analysis when choosing material properties—remembering that linear elastic material constants are inadequate when strains get large. Some programs also have the capability to back calculate material constants from supplied testing data, determining a best fit from either uniaxial or biaxial testing (or both), and some programs have the capability to define the components of the elasticity tensor directly through a user subroutine, allowing complex, nonstandard models to be employed. To assess the performance of a material model implementation, a single element test is often done. The single element test consists of loading a single element and plotting the stress–strain response. This test should be conducted for each material model to ensure that the stress–strain behavior is as expected for the full range of strains used.

Within the load steps of a FEA, loads and boundary conditions are supplied. Each load step allows additional loads to be gradually applied, easing the work of the solver to handle nonlinearities. A collection of load steps allows for consideration of history-dependent phenomena or sequential loadings (such as would be applied to a stent that must be compressed onto a delivery catheter before being deployed on a bench top or within a pulsating artery). Loads in the structural sense can include body forces, point forces, tractions, tensions, and pressures. These loads are applied to geometry (bodies, surfaces, edges, and points) or to aspects of the mesh (nodes, elements, and element faces). Boundary conditions are the controls on possible motions of the geometry or mesh. It is essential that the model be adequately constrained such that a multiplicity of solutions does not exist and rigid body motion is not possible. A multiplicity of solutions exists when the conditions of the model do not prevent one solution from being any less likely or good than another. The solution being referred to is the fully-known state at the end of FEA, where equilibrium is completely described including all values, everywhere, in the model for stiffness, forces, and displacements (and also stresses and strains).

Consider a uniaxial extension test of a dog-bone sample, such as in Figure 9.3. The sample is held fixed in the x direction at one end and stretched by moving the other end away (which is aligned with the x direction), with the sample lying on the x–y plane. The resulting equilibrium state after applying stretch is the solution. The held end has a displacement of $x = 0$, but unless explicitly

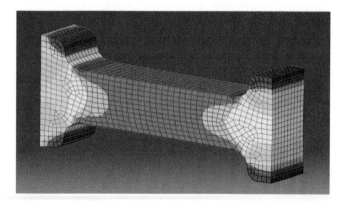

Figure 9.3 Example of dog-bone under displacement-based tension. The figure is color encoded for Von Mises stress, ranging from blue to red—units not shown as analysis parameters chosen for illustration only. Difference in the color pattern on ends occurs because the left end is fixed in x, y, and z, whereas the right end is set to specified x displacement while y and z are unconstrained.

restricted, there is nothing preventing the sample from having a different value of y or z (but stretching just the same amount). To prevent the situation with the sample at $y = 0$ from being as possible as a situation with the sample at $y = 5$ or $y = 50$, the y dimension must also be restricted (typically at the held end). For the same reason, so must $z = 0$ at some point in the model. If one was to rigidly move the entire deformed model, such that the gap between the ends was the same, due to the numeric nature of FEA, it is necessary to have constraints, otherwise the solver cannot handle that the z position is unimportant.

The FEA software will convert the loads and boundary conditions to the nodes during the solver step for each load step. Having the loads and boundary conditions associated with geometry aspects means they do not need to be redefined when the mesh is changed or updated. It is also typically true that in the load steps or outset of the analysis settings is where one selects whether to allow nonlinear geometric dependencies.

Once all the setup of the FEA model has occurred and all attributes of the analysis have been defined, the solver is called to approximate the equilibrium solution where all unknowns are calculated. In a process that typically happens behind the scenes, using the mesh, loads, and boundary conditions, the initial values of the known components of F, K, and u are calculated. Equilibrium will hold for the initial values of these variables both on an element scale and on a model scale. First, F, K, and u are filled in with any specified values for every element. Then, using the nodal connectivity information, global F, global K, and global u can be assembled. This global system of equations can then be solved to determine the value of every unknown component as long as there is adequate information supplied. The global stiffness matrix (K) is often a rather sparse matrix (populated mostly by zeros), and solvers can take advantage of that to avoid inverting the whole matrix. In the case of nonlinear systems, the system cannot usually be solved directly. An iterative approach where a solution is checked and then revised and rechecked before it can be considered correct is used. In strongly nonlinear problems (such as snap through buckling or dynamic loading conditions), these iterative solvers must be properly tuned to function correctly or an explicit solver must be used. An explicit solver applies a one-way solution method that does not go back to check equilibrium at partial loads but instead increments the load in very small steps and proceeds directly from intermediate solutions instead of revising.

Once the solver has completed running or perhaps even as intermediate values are calculated, solutions (i.e., values for every stress, strain, displacement) are stored. Post-processing of the solutions is necessary to interpret the model in light of the original goals. As an example, if the goal was to determine what design would be strong enough, stress can be plotted and areas of high stress needing redesign can be identified. To assess fatigue life, complex post hoc analyses that take advantage of the combination of the simulated results for peak and cyclic stresses can be done. To assess accuracy related to mesh adequacy, models are typically run at increasing mesh densities to ensure that the attributes of interest (e.g., peak stress value and location) are adequately resolved and appropriately consistent. These studies of mesh independence are typically reported and done for the most complex quantify of interest. For example, a stress component is harder than displacement to resolve to the same precision.

9.3 EXAMPLES OF FEA OF BIOMATERIALS—MICROSTRUCTURAL MODELING

FEA can be applied to analyze the biomaterials of a specific implant or tissue, or as a model of the microstructure of the substance or material itself. In these models of a substance's microstructure, the approach involves creating models of each material on a length scale below that typically of interest. That is, if the gross failure or deformation occurs on the mm and cm scales, the material's microstructure is typically modeled at the micron level. This microstructural approach has been used for bone, soft tissue, metals, cartilage, and scaffolds. By modeling at the microstructural level with sufficient geometric understanding and precision, a simplified understanding of a tissue level response of the material can be generated. This approach uses relatively simplistic modeling of the material substance(s), thus saving tremendous computational cost in subsequent models.

9.3.1 Bone

A relatively early example by van Rietbergen et al. utilized FEA based on a high-resolution scan of a portion of bone, along with high-density meshing, to calculate bulk tissue modulus (van Rietbergen et al. 1995). A recent refinement of this work has utilized a more accurate approximation and conversion of the image information into the basis for their model with B-splines and level sets (Verhoosel et al. 2015). Remodeling of bone has also been modeled using the approach of Huiskes et al., and the FEA has been used to calculate updated material properties (Huiskes et al. 2000). Further extensions of this idea have shown that a micro-structurally based approach is more accurate than dual-energy x-ray absorptiometry or bone morphology measurements in predicting Colles-type wrist fractures (Pistoia et al. 2002).

9.3.2 Soft Tissue

Microstructurally motivated models of soft tissues built on FEA are less prevalent, perhaps due to the difficulty in modeling orthotropic and layered structures of many soft tissues, for example, heart valve tissue (Weinberg and Kaazempur-Mofrad 2005).

9.3.3 Metals

Within nitinol, a commonly used flexible implant material, the grains of the metal have been modeled with a microstructural approach as a step toward greater understanding of the fracture behavior of nitinol stents (Weafer and Bruzzi 2014).

9.3.4 Cartilage

To model the mechanical properties of cartilage at a bony interface, such as the hip joint, one group assigned depth-dependent material properties to the cartilage. This approach is based on the inhomogeneous, laminar microstructure of the cartilage as observed in cross section (Osawa et al. 2014).

9.3.5 Scaffolds

Another group used finite elements to optimize the design of their porous scaffold microstructure for bone regeneration. Their model highlighted that complex regeneration behavior can be generated by simple rate equations governing scaffold removal and deposition (Adachi et al. 2006).

9.4 EXAMPLES OF FEA OF BIOMATERIALS—MODELING ANATOMY AND PHYSIOLOGY

FEA can provide insight into the role of biomaterials within normal and pathologic anatomy in a range of systems including within the heart, the heart itself, blood vessels, and orthopedic joints.

The living heart project has built a finite element model of the entire beating human heart that even incorporates an electrical system to represent the autonomic firing of nerves causing contraction of the heart muscle (Baillargeon et al. 2014).

Models of blood vessels include that of Delfino et al. who investigated stresses within the carotid artery (Delfino et al. 1997). The cardiac valve mechanics have been modeled with FEA and other approaches (Fan and Sacks 2014).

Orthopedic finite element models are too numerous to mention and range widely in complexity. One set of models of particular interest is the open knee project, which made the full process behind the development of a series of knee simulations open source. This model has gone on to be the foundation for other groups and studies. The model and related documentation provide a thorough example and grounding in the process of building models from medical images (Erdemir 2016).

9.5 EXAMPLES OF FEA OF BIOMATERIALS—DEVICE EVALUATION

FEA has been applied to biomaterials to represent devices and evaluate designs in isolation as well in virtual use with tissues. Relatively early papers such as the one by Migliavacca utilized FEA to model the response of a stent to loading in relative isolation (Migliavacca et al. 2002). Additional papers also showed the impact of stent design on artery wall mechanics (Prendergast et al. 2003, Bedoya et al. 2006). A paper by Perry et al. further describes how FEA can be used to design cardiovascular stents (Perry et al. 2002).

Models investigating design choices have the potential to expand the understanding of the role mechanics plays in the remodeling of tissues. Specifically, a recent work considered a small change to implant a design that had an effect on induced mechanical stress and found a difference in stress values corresponded with a bone ingrowth response (Chowdhary et al. 2015). The idea that the changes in mechanical stress can precipitate changes in growth patterns has been found in soft tissues as well. Timmins et al. implanted stents of two designs, with one design provoking high stress and the other low in FEA. They observed greater neointimal thickening at stent struts that had higher wall stress in these stents that were implanted in porcine arteries for 28 days (Timmins et al. 2011). These sorts of reactions with living systems are a reminder to the engineer and the designer that the material properties, particularly of living tissues, are not constant but are influenced by their mechanical as well as biochemical environments.

9.6 SUMMARY

Finite element analysis is a prominent tool for the analysis of structures built or designed from biomaterials. Regardless of the program used, a logical progression of steps leads to the calculating of the stresses and strains in one or combined biomaterials. Understanding the assumptions and interpretations of these calculations will lead to a better understanding of the mechanical properties in complex biomaterials and their interactions with host tissues.

REFERENCES

Adachi, T., Y. Osako, M. Tanaka, M. Hojo, and S. J. Hollister (2006). Framework for optimal design of porous scaffold microstructure by computational simulation of bone regeneration. *Biomaterials* **27**(21): 3964–3972.

Baillargeon, B., N. Rebelo, D. D. Fox, R. L. Taylor, and E. Kuhl (2014). The Living Heart Project: A robust and integrative simulator for human heart function. *Eur J Mech A Solids* **48**: 38–47.

Bathe, K. J. (2006). *Finite Element Procedures*. Prentice Hall, USA.

Bedoya, J., C. A. Meyer, L. H. Timmins, M. R. Moreno, and J. E. Moore (2006). Effects of stent design parameters on normal artery wall mechanics. *J Biomech Eng* **128**(5): 757–765.

Chowdhary, R., A. Halldin, R. Jimbo, and A. Wennerberg (2015). Influence of micro threads alteration on osseointegration and primary stability of implants: An FEA and in vivo analysis in rabbits. *Clin Implant Dent Relat Res* **17**(3): 562–569.

Delfino, A., N. Stergiopulos, J. E. Moore, Jr., and J. J. Meister (1997). Residual strain effects on the stress field in a thick wall finite element model of the human carotid bifurcation. *J Biomech* **30**(8): 777–786.

Erdemir, A. (2016). Open knee: Open source modeling and simulation in knee biomechanics. *J Knee Surg* **29**(2): 107–116.

Fan, R. and M. S. Sacks (2014). Simulation of planar soft tissues using a structural constitutive model: Finite element implementation and validation. *J Biomech* **47**(9): 2043–2054.

FDA (2014). *Reporting of Computational Modeling Studies in Medical Device Submissions—Draft Guidance for Industry and Food and Drug Administration Staff*, Center for Devices & Radiological Health, U.S. Food & Drug Administration.

Hayenga, H. N., S. Kubecka, A. Morris, R. Bhui, and C. A. Meyer (2014). A mechanobiological model of atherogenesis. *World Congress of Biomechanics, Boston, MA.*

Huiskes, R., R. Ruimerman, G. H. Van Lenthe, and J. D. Janssen (2000). Effects of mechanical forces on maintenance and adaptation of form in trabecular bone. *Nature* **405**(6787): 704–706.

Kwon, Y. W. and H. Bang (2000). *The Finite Element Method Using MATLAB*, 2nd edn. CRC Press, Boca Raton, FL.

MacDonald, B. J. (2011). *Practical Stress Analysis with Finite Elements*. Glasnevin Publishing, Dublin, Ireland.

Migliavacca, F., L. Petrini, M. Colombo, F. Auricchio, and R. Pietrabissa (2002). Mechanical behavior of coronary stents investigated through the finite element method. *J Biomech* **35**(6): 803–811.

Osawa, T., S. Moriyama, and M. Tanaka (2014). Finite element analysis of hip joint cartilage reproduced from real bone surface geometry based on 3D-CT image. *JBSE* **9**(2): 13-00164–113-00164.

Pathmanathan, P. and R. A. Gray (2013). Ensuring reliability of safety-critical clinical applications of computational cardiac models. *Front Physiol* **4**: 358.

Perry, M., S. Oktay, and J. C. Muskivitch (2002). Finite element analysis and fatigue of stents. *Minim Invas Therapy Allied Technol* **11**(4): 165–171.

Pistoia, W., B. van Rietbergen, E. M. Lochmüller, C. A. Lill, F. Eckstein, and P. Rüegsegger (2002). Estimation of distal radius failure load with micro-finite element analysis models based on three-dimensional peripheral quantitative computed tomography images. *Bone* **30**(6): 842–848.

Prendergast, P. J., C. Lally, S. Daly, A. J. Reid, T. C. Lee, D. Quinn, and F. Dolan (2003). Analysis of prolapse in cardiovascular stents: A constitutive equation for vascular tissue and finite-element modelling. *J Biomech Eng* **125**(5): 692–699.

Reddy, J. N. (2015). *An Introduction to Nonlinear Finite Element Analysis: With Applications to Heat Transfer, Fluid Mechanics, and Solid Mechanics*. Oxford University Press, New York, NY.

Timmins, L. H., M. W. Miller, F. J. Clubb, Jr., and J. E. Moore, Jr. (2011). Increased artery wall stress post-stenting leads to greater intimal thickening. *Lab Invest* **91**(6): 955–967.

van Rietbergen, B., H. Weinans, R. Huiskes, and A. Odgaard (1995). A new method to determine trabecular bone elastic properties and loading using micro-mechanical finite-element models. *J Biomech* **28**(1): 69–81.

Verhoosel, C. V., G. J. van Zwieten, B. van Rietbergen, and R. de Borst (2015). Image-based goal-oriented adaptive isogeometric analysis with application to the micro-mechanical modeling of trabecular bone. *Comput Methods Appl Mech Eng* **284**: 138–164.

Weafer, F. M. and M. S. Bruzzi (2014). Influence of microstructure on the performance of nitinol: A computational analysis. *J Mater Eng Perform* **23**(7): 2539–2544.

Weinberg, E. J. and M. R. Kaazempur-Mofrad (2005). On the constitutive models for heart valve leaflet mechanics. *Cardiovasc Eng* **5**(1): 37–43.

PART IV
BIOMATERIAL PERSPECTIVES

10 Perspectives on the Mechanics of Biomaterials in Medical Devices

Heather N. Hayenga and Kim L. Hayenga

CONTENTS

10.1 ADVANCEMENT OF BIOMATERIALS

Considerable advancements in the understanding and the fabrication of biomaterials for medical devices have been made over the years, yet the field is still in its infancy. Historically, medical implant designs were motivated solely by the need to restore function to a patient. Early medical devices were made from available materials, such as ivory, bone, and wood, and the skill of the resident surgeon in order to replace lost or damaged limbs. Over time, medical devices utilized metals (e.g., gold, silver, or amalgams in dentistry), natural materials (e.g., porcine heart valves in humans), alloys (e.g., surgical stainless steel in orthopedic implants), and polymers (e.g., contact lenses). A majority of the engineered materials used in medical devices have occurred in the last 60 years. This has led to the growing research field of biomaterial sciences and engineering in an attempt to design, synthesize, characterize, and implant the optimal biomimetic material for a particular disease or replace the function of a tissue or organ.

10.2 EXAMPLE: EVOLUTION OF VASCULAR STENTS

One example of a medical device that has seen major refinements in its design and composition over the last 30 years is the vascular coronary stent/scaffold. A vascular stent/scaffold is needed to reopen and brace an artery after it has undergone severe narrowing or complete occlusion as in the case after an advanced atherosclerotic plaque develops in the artery wall. Table 10.1 shows the evolution of design, material, dimension, and mechanical properties used for vascular stents/scaffolds. In the early 1980s, investigators used canines to test stents of different materials and designs [1]. For example, Dotter et al. developed a coiled nitinol "memory metal" that expanded to a predetermined diameter upon the injection of heated saline to the blood vessel [2] and Palmaz et al. developed expandable stainless steel tubular mesh stents that could be mounted on angioplasty catheters with an outer constraining sheath [3]. These and other preclinical studies were met with enough success that the first stainless steel stent was placed in a human coronary artery in 1986 and led to the development of many other bare metal stents (BMSs) in the 1990s [4,5]. Although these stents restored blood flow, there were drawbacks. The ridged metal encouraged subacute stent thrombosis, and long-term restenosis (or re-narrowing of the lumen due to increased cellular proliferation within the artery wall). To address these limitations, a polymer containing anti-thrombogenic and anti-proliferation drugs

Table 10.1: Overview of the Evolution of Cardiovascular Stents

Stent Type	BMS—Cordis Palmaz-Schatz (P-S 153)	BMS—Medtronic BeStent Brava	DES—Boston Scientific PROMUS Element Plus	BRS—Biotronik AMS-1
Material(s)	Stainless steel	Stainless steel	Platinum–chromium	93% Magnesium and 7% rare earth metals
Strut thickness (μm)	100	100	81	165
Polymer thickness (μm)	NA	NA	3	NA
Elastic recoil (%)	3.8	2.18	3	<8
Collapse pressure	0.5 bar	1.0 bar	NA	0.8 bar
Axial shortening (%)	4	3.7	2.3 ± 6.0	<5
References	[25–28]	[29,30]	[31,32]	[5,17,33]

Notes: Representation of select BMS, DES, and bioresorbable scaffolds detailing their material(s), strut thickness, and mechanical properties. Notably, the BRS are thicker and exhibit more elastic recoil (or compress more in the radial direction after implant) than the BMS and DES. The collapse pressure, or maximum pressure of the artery the stent can withstand before it collapses varies for each stent; stents need to have a minimum collapse pressure above 0.4 bar (300 mmHg) in order to keep the artery open [34]. The axial shortening is comparable between all stent types.

Abbreviations: BRS, bioresorbable scaffold; NA, not applicable (either not relevant or values are not in the public domain).

was formulated and applied to the metal stents creating the next generation of stents, drug-eluting stents (DESs).

In 1999, Eduardo Sousa et al. implanted the first DES into a human coronary artery [6]. This stent eluted the immunosuppressive compound sirolimus; this fungus-derived compound targets a receptor that leads to cessation of cell-cycle progression and hence inhibits vascular smooth muscle cell (VSMC) proliferation [7]. However, as with the BMS, the enthusiasm was again curbed when studies were linking DES to an increased risk of stent thrombosis. The increased risk of stent thrombosis may be due to delayed endothelialization by anti-proliferation drugs or delayed immune reaction to the polymer in DES [8–11]. Anti-platelet therapy for 6–12 months has been successful in reducing major adverse cardio-vascular events (MACEs) compared to the placebo [5]. However, since the BMS and DES have advantages and disadvantages that are patient-specific, both stents continue to be supplied. Yet the concerns of the polymer leading to a long-term immune reaction and thrombosis provided the incentive to develop stents that provide only a temporary scaffold, that is, biodegradable stents.

The newest development used to open the vessel wall is the biodegradable scaffold. A poly-l-lactic acid (PLLA) polymer scaffold will open the vessel and over time degrade. In 1998, Hideo Tamai et al. implanted the first fully biode-gradable PLLA coronary stent in a human [12]. The results of a 10-year follow-up of the initial 50 patients to receive this bioresorbable scaffold (BRS) were decent, with 98% surviving free of cardiac death, 50% had a major adverse cardiac event, and 38% had restenosis of the target vessel [13]. The scaffold had a helical zigzag design, which did not injure the vessel wall as much as the previous knitted patterns and therefore reduced thrombus and intimal hyperplasia [12]. PLLA is a frequent material used for BRS, as well as other health care devices (e.g., degrad-able sutures, soft-tissue implants, and orthopedic implants). An appropriate stent material needs to have a high-elastic moduli to impart radial stiffness, high-break strain to withstand deformations from the crimped to expanded states, and low-yield strains to reduce the amount of recoil and overinflation necessary to achieve a target deployment. In order to withstand higher strains, PLLA and other BRS on the market needed to increase the average stent strut thickness to about 160 μm as compared to the average DES strut thickness of only 104 μm [5]. The time course of degradation itself depends on various factors including the chemical bond, the pH, the presence of catalysts, and the co-polymer composi-tion. Polymer biodegradation is generally a hydrolytic process starting with the penetration of water into the polymer, hydrolysis of the ester bonds between repeating lactide units, and eventually the conversion to carbon dioxide and water. The Igaki-Tamai BRS typically takes about 2 years to fully reabsorb.

Since the introduction of the first BMS, DES, and BRS, they have under-gone many refinements in their material and design. For example, BMSs were initially made of stainless steel, in part because this material is biologically inert. However, stainless steel was superseded by cobalt–chromium or platinum–chromium alloys as they can be designed significantly thinner (without compromising radial strength or corrosion resistance) and have higher radio-opacity and conformability [14]. In addition, DES incorporated these materials, and nickel–titanium (nitinol) as their platform allowing for thinner struts and them to be more biologically inert. The stent material influences the mechani-cal behavior of the stent, but since only the outer few micrometers of the metal interact with the blood and tissue wall, DES advanced by incorporating new and promising drug combinations for anti-inflammatory, anti-thrombogenic, anti-proliferative, and immunosuppressive properties. DES also advanced by incorporating novel polymers with optimal physical properties, stability, compatibility with drugs, biocompatibility with vascular tissue, and control of drug release [15]. After the first generation of permeate (i.e., nondegradable) polymer coatings (e.g., polyethylene-co-vinyl acetate, poly-*n*-butyl methacrylate, and poly(styrene-*b*-isobutylene-*b*-styrene)), more biocompatible but still permanent polymers (e.g., phosphorylcholine (a constituent of the lipid bilayer of the cell membrane) and hyaluronic acid) were used to coat DES. These polymers provided a hydrophilic surface while retaining the overall neutral charge [16]. Now, DESs contain bioabsorbable polymers (e.g., polyglycolic or polylactic acid) that degrade over time. The idea being the polymer and drug will have controlled biodegradation and release, respectively, so that only the stent platform (BMS) will be left behind in the future; thus, minimizing any harmful effects of the polymer. However, completely degradable stents would obviate the need for long-term anti-platelet therapy since no foreign material would be left behind. Moreover, future surgical options will not be limited and follow-up imaging such as MRI would be possible.

The current BRSs are composed of either a polymer or bioresorbable metal alloy. Numerous different polymers are available, each with different chemical compositions, mechanical properties, and subsequently bioabsorption times. The first approved BRS in the United States, Absorb scaffold (Abbott Vascular, Santa Clara, CA), has already undergone refinements to enhance the mechanical strength of struts and to reduce early and late recoil. Specifically, the new design has zigzag hoops linked by bridges to reduce the unsupported surface area due to radial recoil. The manufacturing process for the Absorb scaffold was also modified so that the polymer degraded (hydrolyzed) slower in vivo, thus preserving its mechanical integrity for a longer period of time [17,18]. In addition to polymer BRS, the first metal absorbable stent (AMS-1; Biotronik, Berlin, Germany) was implanted on July 30, 2004. This metal BRS was 93% magnesium; magnesium is the fourth most common cation in the human body and is responsible for the synthesis of over 300 enzymes, cofactor for ATPase activation, vasodilation, stent thrombosis inhibitor (due to electronegative charge) [19], and promotion of the recruitment of collaterals during ischemia. Initially, the mechanical strength and properties of these stents were comparable to those of stainless steel stents (high collapse pressure of 0.8 bar, minimal shortening after inflation <5%, and low elastic recoil <8%). However, early magnesium BRS (AMS-1) degraded too rapidly to prevent constrictive remodeling and restenosis. In one patient, 50% of the struts were degraded after only 3 weeks. Stent degradation compromises mechanical properties making the stent unable to resist the elastic recoil or inward force imposed by the artery. Consequently, the restenosis rate at 4 months was nearly 50%. Although tremendous progress has been made to improve the material and design of coronary stents, there is still room for continued improvement.

Future generations of stents may allow the artery to recoil, as opposed to the ridged stents today, thereby preserving the functionality of the artery. The elastic recoil of the artery happens after each systolic pump of the heart and is needed to "push" the blood downstream to the extremities. This elastic recoil generates cyclic strain, typically 0%–10%, on the artery wall. The strain in turn is experienced by VSMCs and affects their behavior. For example, in conditions of high strain, VSMCs synthesize more platelet-derived growth factor, which results in increased proliferation [20–22]. Moreover, under high strains, VSMCs produce more tissue growth factor and matrix metalloproteinases leading to the production and reorganization of collagen. Therefore, to restore the natural recoil ability of a coronary artery and therefore minimize restenosis, the deployed stent should provide the needed support while being flexible enough to not increase the stain experienced by VSMCs. Indeed, experimental studies have shown that flexible stents reduce the neointimal proliferation compared to ridged stents [23]. However, the flexible stents were not able to maintain large diameters in the long term; reinforcing a need for balance between the mechanical support and biological response of the stent material.

Thus, the next generation of stents may be mostly bioabsorbable with a thin structure of metal. If made of the right material (e.g., graphene, cobalt–platinum), the struts can be manufactured thinner while maintaining the structural support needed. The thin strut dimensions (below 75 μm) should ensure rapid endothelialization on the implant. Having an intact endothelium should help reduce neointimal proliferation. For example, a BMS with thin strut thickness of 65 μm (SolarFlex cobalt–chromium stent) was implanted in 240 patients in 2011 and showed <6% MACE at 6 months [24]. A drug-eluting bioabsorbable polymer coating should reduce the immune response in the beginning.

10.3 PHYSICAL AND CHEMICAL CONSIDERATIONS OF BIOMATERIALS IN MEDICAL DEVICES

The fields of biomaterials and mechanobiology are still in their initial stages despite considerable advancements in understanding how the physical properties of materials affect biological functions. It is clear that physical properties, such as size, shape, mechanical properties, surface texture, density, porosity, and compartmentalization, profoundly impact the function of a biomaterial. This greatly widens the design parameter space for the next generation of biomaterials but, at the same time, raises important questions. To what degree does the biological response depend on a specific physical property of the material? What is the relative weight of different physical and chemical factors on the biological response? For each biomedical application, the mechanisms of how physical properties affect biological performance, as well as the interplay between various physio-mechano-chemical properties, may have to be elucidated case by case.

Over the past decade, the manufacturing processes of several materials have been refined to control their physical properties. The underpinning processing conditions should allow for sensitivity to elevated temperatures, organic solvents, or extreme pH changes. The ability to multiplex physical and chemical design parameters amongst single and multiple materials will help ensure compatibility and optimal performance [35]. For example, rational combinations of shape, size, flexibility, and surface chemistry have been considered to create particles that can target select vascular endothelium and effectively deliver necessary drugs [36]. Drug delivery applications open the multiparameter toolbox researchers can explore in order to synthesize the perfect particle. For example, size affects clearance, that is, the removal of liposomes by mononuclear phagocytes increases with size for particles greater than 100 nm [37]. Whereas for liposomes less than 100 nm, charge may be more important; positive charges are thought to help the particle escape the endosomes of cells [38]. Moreover, particle flexibility and surface modification with polyethylene glycol can prolong the presence of the particles in the blood circulation [36]. It is an exciting time to explore the various parameters that control the physio-mechano-chemical properties of biomaterials for drug-delivery applications.

For many biomedical implants, there is a need to find the ideal balance in mechanical behavior. That is, the biomaterial may need to be stiff enough to meet the load demands and serve its purpose yet soft enough to exhibit biomimicry and not elicited an adverse host response. Consider a vascular stent, if the stent was made purely on biomimicry the structure would collapse immediately after implant. Conversely, if the stent overstretches the artery greater than 20%, permanent damage to the arterial constituents will result [39]. A stent that is too stiff will increase the chance of restenosis and the inflammatory response [23,40]. Advances in material synthesis have allowed for novel thin-film nitinol materials that can withstand loads while being flexible enough to undergo deformations during pulsatility, stretching (as in the case of some peripheral arteries), or trauma [41,42]. However, the challenge remains in making these materials also biodegradable. In many situations, it is desirable for the material to naturally degrade over time to allow the natural tissue to grow through it and repair itself. However, degradable materials are limited mostly to hydrolytically degradable polymers and many suffer from not being able to withstand significant loads. Hydrolytically degradable polymers are materials that possess chemical bonds in their backbone that can easily be broken yielding two species, one gaining a hydrogen group and another a hydroxyl group. See Ulery et al. for a nice review of the number of hydrolytically sensitive polymeric families, degradation rates, and their biomedical applications [43].

Modeling techniques and experimental characterization can improve the mechanical demands we should be targeting. As mentioned earlier in this book in Chapter 7,

> "computational biomechanical models show great promise in advancing the field of vascular tissue engineering. They enable time- and cost-efficient evaluations of fundamental hypotheses to reduce the experimental search space, and aid investigators in moving away from a purely empirical, trial-and-error approach toward rational design."

In this work, the authors used a parameter sensitivity approach to optimize the values for four primary classes of parameters in vascular graft design: scaffold and matrix stiffness, scaffold geometry (e.g., porosity, fiber diameter, pore size), scaffold and matrix degradation rates, and matrix production rates (inflammation- and mechano-mediated). In addition, modeling can be used on a smaller scale to determine the molecular physicochemical interactions in biological systems and between biomaterials as discussed in more detail in Chapter 8. Overall inclusive models that apply principles from engineering, material science, and biology will help develop and understand the mechanics of new materials better.

10.4 NOVEL BIOMATERIALS

New fabrication techniques have led to the design of "smart" materials. Smart or reactionary materials can significantly change one or more of its properties in reaction to an external stimulus such as stress, temperature, pH, and electric or magnetic fields. These changes typically occur in a dependable, reproducible, and usually reversible manner. For example, investigators have synthesized polymers that are able to switch their mechanical properties between soft and stiff, much like a sea cucumber dermis does when threatened. Specifically, intercortical microelectrodes were developed that can be sufficiently stiff to penetrate the tissue upon implantation, but once inside the hydrated brain the electrodes soften to better match the mechanical properties of the cortical tissue [44,45]. This softening effect can be achieved by reinforcing the polymer (poly(vinyl acetate)) with rigid cellulose nanocrystals that disassemble upon exposure to fluid, bringing the stiffness from 5.1 GPa down to 12 MPa. However, this requires the polymer to swell considerably (30%–90% w/w) [46]. More recently, investigators have developed shape memory polymer (SMP) microelectrodes from acrylate and thiol-ene/acrylate polymers that soften from over 1 GPa to 18 MPa after insertion but swells less than 3% [47,48]. Less water accumulation into the biomaterial may prevent delamination and increase the compatibility of the biomaterial with its environment. Although an improvement, there is still room for improvement as the stiffness of the cortical brain tissue is in the order of 1 kPa and not MPa [49]. Smart materials have been able to reduce the mechanical mismatch of the implanted device to the tissue, which may reduce the neuro-inflammatory response [45]. Thus, smart materials have the potential to overcome current limitations of traditional materials through their dynamic interaction with the environment.

In addition to smart materials, 3D-printing, a convenient way to create materials with custom shapes and designs, has received attention recently. The Food and Drug Administration (FDA) has already cleared more than 85 3D-printed medical devices (as of December 2015; FDA.gov). So far these medical devices have been either bone-related (orthopedic and cranial) implants or external (surgical instrument, dental restorations such as crowns, and external prosthetics). Senatov et al. have recently used 3D-printing to make a shape memory polylactide and hydroxyapatite porous polymer with robust mechanical properties and designed for osteointegration and conformity within cracks of bone [50]. Moreover, a recent clinical trial showed that 3D-printed osteosynthesis plates performed better than conventional plates in treating intercondylar humeral fractures [51].

Another type of biomaterial that has significantly advanced in the past few decades is the nanomaterial for drug delivery. The nanomaterial should target the desired tumor location and not cause unwanted side effects. Optimum size and shape help achieve this desired goal. For example, too small (radius < 10 nm) and hydrophobic polymers tend to get absorbed by endothelial cells lining the blood vessels or filtered by kidneys more rapidly than hydrophilic polymers with a slightly larger and cylindrical shape [52]. Making these two changes to a nanomaterial can extend the time in circulation from 2 to >100 h [53]. Since tumor vasculature is typically much "leakier" to colloidal particles than the healthy vasculature, traditional targeting techniques have been used to keep the nanomaterial in the circulation longer and hence increasing the amount that will eventually accumulate in the tumor. Within the last decade, researchers have developed nanomaterials that can target unique molecular features on the tumor endothelium (e.g., annexin 1 ligand) [54,55], penetrate through cell membranes to deliver genetic material (e.g., siRNA) [56], transport to specific regions via an applied magnetic field [57], and even enter a cell and travel close to the nucleus before releasing DNA-targeting anticancer drugs (i.e., pH-responsive nanoparticles) [58]. These examples illustrate how multicomponent polymeric nanomaterials can guide drugs to specific tissues, coax them through biological barriers within cells, and even escape premature clearance. Nevertheless, the next challenge is to ensure these nanomaterials maintain their target specificity, stability, size, and proper clearance in the complex protein environment of the body.

10.5 CONCLUSION

In conclusion, new processing techniques have enabled the fabrication of biomaterials that are more biomimetic than ever before. It can even be argued that some biomaterials have surpassed purely biomimetic materials. By engineering vascular stents/scaffolds with different designs and materials, the stents have been able to reduce strut thickness or reduce foreign body response by reabsorbing, thus reducing the rate of adverse events overall. Moreover, advanced manufacturing processes and computational modeling have allowed for a better understanding of how physical properties affect biological performance, as well as the interplay between various physio-mechano-chemical properties. There are many novel biomaterials on the horizon including smart materials, SMPs, 3D-printed biomaterials, and nanomaterials that have the potential to revolutionize health care in the future.

REFERENCES

1. Ruygrok, P.N. and P.W. Serruys, Intracoronary stenting. From concept to custom. *Circulation*, 1996. **94**(5): 882–890.

2. Dotter, C.T. et al., Transluminal expandable nitinol coil stent grafting: Preliminary report. *Radiology*, 1983. **147**(1): 259–260.

3. Palmaz, J.C. et al., Expandable intraluminal graft: A preliminary study. Work in progress. *Radiology*, 1985. **156**(1): 73–77.

4. Sigwart, U. et al., Intravascular stents to prevent occlusion and restenosis after transluminal angioplasty. *N Engl J Med*, 1987. **316**(12): 701–706.

5. Iqbal, J., J. Gunn, and P.W. Serruys, Coronary stents: Historical development, current status and future directions. *Br Med Bull*, 2013. **106**: 193–211.

6. Sousa, J.E., P.W. Serruys, and M.A. Costa, New frontiers in cardiology: Drug-eluting stents: Part I. *Circulation*, 2003. **107**(17): 2274–2279.

7. Poon, M. et al., Rapamycin inhibits vascular smooth muscle cell migration. *J Clin Invest*, 1996. **98**(10): 2277–2283.

8. Camenzind, E., P.G. Steg, and W. Wijns, Stent thrombosis late after implantation of first-generation drug-eluting stents: A cause for concern. *Circulation*, 2007. **115**(11): 1440–1455; discussion 1455.

9. McFadden, E.P. et al., Late thrombosis in drug-eluting coronary stents after discontinuation of antiplatelet therapy. *Lancet*, 2004. **364**(9444): 1519–1521.

10. Hofma, S.H. et al., Indication of long-term endothelial dysfunction after sirolimus-eluting stent implantation. *Eur Heart J*, 2006. **27**(2): 166–170.

11. Joner, M. et al., Pathology of drug-eluting stents in humans: Delayed healing and late thrombotic risk. *J Am Coll Cardiol*, 2006. **48**(1): 193–202.

12. Tamai, H. et al., Initial and 6-month results of biodegradable poly-L-lactic acid coronary stents in humans. *Circulation*, 2000. **102**(4): 399–404.

13. Nishio, S. et al., Long-term (>10 years) clinical outcomes of first-in-human biodegradable poly-L-lactic acid coronary stents: Igaki-Tamai stents. *Circulation*, 2012. **125**(19): 2343–2353.

14. Garg, S. and P.W. Serruys, Coronary stents: Current status. *J Am Coll Cardiol*, 2010. **56**(10 Suppl): S1–S42.

15. Khan, W., S. Farah, and A.J. Domb, Drug eluting stents: Developments and current status. *J Control Release*, 2012. **161**(2): 703–712.

16. Whelan, D.M. et al., Biocompatibility of phosphorylcholine coated stents in normal porcine coronary arteries. *Heart*, 2000. **83**(3): 338–345.

17. Onuma, Y. and P.W. Serruys, Bioresorbable scaffold: The advent of a new era in percutaneous coronary and peripheral revascularization? *Circulation*, 2011. **123**(7): 779–797.

18. Capodanno, D., F. Dipasqua, and C. Tamburino, Novel drug-eluting stents in the treatment of de novo coronary lesions. *Vasc Health Risk Manag*, 2011. **7**: 103–118.

19. Heublein, B. et al., Biocorrosion of magnesium alloys: A new principle in cardiovascular implant technology? *Heart*, 2003. **89**(6): 651–656.

20. Li, Z., S. Moore, and M.Z. Alavi, Mitogenic factors released from smooth muscle cells are responsible for neointimal cell proliferation after balloon catheter deendothelialization. *Exp Mol Pathol*, 1995. **63**(2): 77–86.

21. Ma, Y.H., S. Ling, and H.E. Ives, Mechanical strain increases PDGF-B and PDGF beta receptor expression in vascular smooth muscle cells. *Biochem Biophys Res Commun*, 1999. **265**(2): 606–610.

22. Thorne, B.C. et al., Toward a multi-scale computational model of arterial adaptation in hypertension: Verification of a multi-cell agent based model. *Front Physiol*, 2011. **2**: 20.

23. Fontaine, A.B. et al., Stent-induced intimal hyperplasia: Are there fundamental differences between flexible and rigid stent designs? *J Vasc Interv Radiol*, 1994. **5**(5): 739–744.

24. Suttorp, M.J. et al., Ultra-thin strut cobalt chromium bare metal stent usage in a complex real-world setting. (SOLSTICE registry). *Netherlands Heart J*, 2015. **23**(2): 124–129.

25. Schrader, S.C. and R. Beyar, Evaluation of the compressive mechanical properties of endoluminal metal stents. *Cathet Cardiovasc Diagn*, 1998. **44**(2): 179–187.

26. Kiousis, D.E., A.R. Wulff, and G.A. Holzapfel, Experimental studies and numerical analysis of the inflation and interaction of vascular balloon catheter-stent systems. *Ann Biomed Eng*, 2009. **37**(2): 315–330.

27. Rieu, R. et al., Radial force of coronary stents: A comparative analysis. *Catheter Cardiovasc Interv*, 1999. **46**(3): 380–391.

28. Early, M. and D.J. Kelly, The role of vessel geometry and material properties on the mechanics of stenting in the coronary and peripheral arteries. *Proc Inst Mech Eng H*, 2010. **224**(3): 465–476.

29. Schmitz, K.-P. et al., In-vitro examination of clinically relevant stent parameters. *Prog Biomed Res*, 2000. **5**: 197–203.

30. Engin, A.E. and P.F. Niederer, *Technology and Health Care*, Vol. **9**. 2001: *IOS Press*, Amsterdam, Netherlands, p. 228.

31. Muramatsu, T. et al., Progress in treatment by percutaneous coronary intervention: The stent of the future. *Rev Esp Cardiol (Engl Ed)*, 2013. **66**(6): 483–496.

32. Dvir, D. et al., Abstract 18645: Axial integrity of coronary stents: Evaluation using intravascular ultrasound. *Circulation*, 2012. **126**(Suppl 21): A18645.

33. Erbel, R. et al., Temporary scaffolding of coronary arteries with bioabsorbable magnesium stents: A prospective, non-randomised multicentre trial. *Lancet*, 2007. **369**(9576): 1869–1875.

34. Baer, G.M. et al., Thermomechanical properties, collapse pressure, and expansion of shape memory polymer neurovascular stent prototypes. *J Biomed Mater Res B Appl Biomater*, 2009. **90**(1): 421–429.

35. Yoshida, M. and J. Lahann, Smart nanomaterials. *ACS Nano*, 2008. **2**(6): 1101–1107.

36. Muzykantov, V. and S. Muro, Targeting delivery of drugs in the vascular system. *Int J Transp Phenom*, 2011. **12**(1–2): 41–49.

37. De Jong, W.H. and P.J. Borm, Drug delivery and nanoparticles: Applications and hazards. *Int J Nanomed*, 2008. **3**(2): 133–149.

38. Panyam, J. et al., Rapid endo-lysosomal escape of poly(DL-lactide-co-glycolide) nanoparticles: Implications for drug and gene delivery. *FASEB J*, 2002. **16**(10): 1217–1226.

39. Bell, E.D., J.W. Sullivan, and K.L. Monson, Subfailure overstretch induces persistent changes in the passive mechanical response of cerebral arteries. *Front Bioeng Biotechnol*, 2015. **3**: 2.

40. Hayenga, H.N. and H. Aranda-Espinoza, Stiffness increases mononuclear cell transendothelial migration. *Cell Mol Bioeng*, 2013. **6**(3): 253–265.

41. Nishi, M. et al., Popliteal artery aneurysm treated with implantation of a covered stent graft (fluency) reinforced with a nitinol stent (S.M.A.R.T.). *Cardiovasc Interv Ther*, 2016. **31**(4): 316–320.

42. Shayan, M. and Y. Chun, An overview of thin film nitinol endovascular devices. *Acta Biomater*, 2015. **21**: 20–34.

43. Ulery, B.D., L.S. Nair, and C.T. Laurencin, Biomedical applications of biodegradable polymers. *J Polym Sci B Polym Phys*, 2011. **49**(12): 832–864.

44. Ware, T. et al., Thiol-click chemistries for responsive neural interfaces. *Macromol Biosci*, 2013. **13**(12): 1640–1647.

45. Jorfi, M. et al., Progress towards biocompatible intracortical microelectrodes for neural interfacing applications. *J Neural Eng*, 2015. **12**(1): 011001.

46. Shanmuganathan, K. et al., Stimuli-responsive mechanically adaptive polymer nanocomposites. *ACS Appl Mater Interfaces*, 2010. **2**(1): 165–174.

47. Simon, D. et al., A comparison of polymer substrates for photolithographic processing of flexible bioelectronics. *Biomed Microdev*, 2013. **15**(6): 925–939.

48. Ware, T. et al., Thiol-ene/acrylate substrates for softening intracortical electrodes. *J Biomed Mater Res B Appl Biomater*, 2014. **102**(1): 1–11.

49. van Dommelen, J.A. et al., Mechanical properties of brain tissue by indentation: Interregional variation. *J Mech Behav Biomed Mater*, 2010. **3**(2): 158–166.

50. Senatov, F.S. et al., Mechanical properties and shape memory effect of 3D-printed PLA-based porous scaffolds. *J Mech Behav Biomed Mater*, 2016. **57**: 139–148.

51. Shuang, F. et al., Treatment of intercondylar humeral fractures with 3D-printed osteosynthesis plates. *Medicine (Baltimore)*, 2016. **95**(3): e2461.

52. Christian, D.A. et al., Flexible filaments for in vivo imaging and delivery: Persistent circulation of filomicelles opens the dosage window for sustained tumor shrinkage. *Mol Pharm*, 2009. **6**(5): 1343–1352.

53. Hubbell, J.A. and A. Chilkoti, Chemistry. Nanomaterials for drug delivery. *Science*, 2012. **337**(6092): 303–305.

54. von Maltzahn, G. et al., Nanoparticles that communicate in vivo to amplify tumour targeting. *Nat Mater*, 2011. **10**(7): 545–552.

55. Hatakeyama, S. et al., Targeted drug delivery to tumor vasculature by a carbohydrate mimetic peptide. *Proc Natl Acad Sci USA*, 2011. **108**(49): 19587–19592.

56. Davis, M.E. et al., Evidence of RNAi in humans from systemically administered siRNA via targeted nanoparticles. *Nature*, 2010. **464**(7291): 1067–1070.

57. Mody, V.V. et al., Magnetic nanoparticle drug delivery systems for targeting tumor. *Appl Nanosci*, 2014. **4**(4): 385–392.

58. Murakami, M. et al., Improving drug potency and efficacy by nanocarrier-mediated subcellular targeting. *Sci Transl Med*, 2011. **3**(64): 64ra2.

11 A Perspective on the Impact of Additive Manufacturing on Future Biomaterials

Jesse K. Placone and John P. Fisher

CONTENT

The development of biomaterials has seen many rapid advances and changes in approach over the past decade. To this end, additive manufacturing, or 3D printing, has been a major driving force in the development of novel biomaterials and applications of these biomaterials. These recent advances have been largely brought about by the combination of multiple approaches used in the fields of engineering, materials science, and stem cell biology to model, create, and understand the underlying fundamentals that govern the invasion/recruitment, growth, expansion, and differentiation of desired cell types in the biomaterial [1–4]. Despite these advances, there is still a need for systematic approaches for the development of novel biomaterials in which their physical properties are well understood and their interactions with the desired cell types are modeled and controlled prior to undertaking large, expensive studies investigating multiple variables. We envision that the future of biomaterial development will focus on utilizing a combination of natural and synthetic materials to better mimic native tissue and will utilize multiple cell types to recapitulate the *in vivo* niche environment to facilitate self-assembly within the biomaterial.

Through the use of 3D printing and 3D bioprinting, the size of scaffolds that can be readily fabricated is steadily increasing [5,6]. With this increase in size and complexity, there is a corresponding increase in the metabolic demands placed upon the scaffold due to the increase in cell numbers [7]. Recent work has focused on fabricating engineered tissues that are no longer single materials but are multi-material composites to address this increased need for nutrient exchange *in vitro* and *in vivo* [5,8]. This complexity has driven the need for more complex in silico modeling of the biomaterial to ensure adequate nutrient and oxygen levels as well as to inform the levels of growth factor loading to help guide stem cell fate within the scaffolds. Nutrient exchange can be driven by prefabricated vasculature, the inherent porosity of the biomaterial, or through a combination of the two under dynamic (e.g., bioreactor culture) or static culture [8].

The development of future biomaterials should not be limited to basic materials but should make use of composites. For example, 3D bioprinting allows for the concurrent fabrication of multicomponent interfacial tissues/biomaterials. This can be seen as advantageous toward the development of many engineered tissues such as bone, cartilage, skeletal muscle, skin, vasculature, and organogenesis in general [4–6,9–15]. As discussed in this book, there are many techniques to address this need for increased complexity while developing materials that take cues from nature. One method that will see a much more widespread adoption may be the use of decellularized tissues. These biomaterials contain natural binding domains, structural features, and cues for the enhanced response of stem cells and desired cell types. As such, these materials when utilized with 3D printing show great promise for advancing our understanding of biomaterial–cell interactions.

Additive manufacturing allows for the spatial control of the resultant biomaterial. One critical aspect of this for the future development of biomaterials is the ability to spatially control the deposition of the material as well as its physical properties (e.g., stiffness, permeability, viscoelastic properties, swelling, as well as others) [10,16–18]. By modulating these properties, the resultant biomaterial can be tailored to better mimic native tissue. As an example application, there has been a recent push toward developing biomaterials suitable for organogenesis. These biomaterials have provided unique insight into development pathways and have established parameters that are necessary for the generation of complex tissues with stem cells. Future applications of these biomaterials and resultant organoids are not limited to understanding developmental biology, but they can be utilized to enhance treatment options in regenerative medicine. One such application currently under investigation is the development of organoids for drug screening and developmental biology [9]. These biomaterial systems consisting of hydrogels laden with multiple cell types, derived from pluripotent stem cell sources, will potentially allow for the assessment of drug toxicity, efficacy, and systemic effects on models for human organs *in vitro* [6]. Thereby, the drug testing and development process can be transitioned to a more high-throughput process.

Through incorporating growth factors and other cytokines and controlling their time-dependent release [19], future biomaterials will potentially address multiple challenges in the field of tissue engineering by recruiting specific cell types to confined areas and controlling their fate. Furthermore, it has been well established that cell–cell signaling is critical for the organization, function, and homeostasis of the developing tissue [3]. Future biomaterials need to address this critical need by tightly controlling the spatial placement of individual cell populations, thus controlling their interactions, mediated through direct cell–cell contact, paracrine, or other signaling mechanisms. As such, biomaterials will begin to better mimic the native structure and organization of natural tissues, albeit initially on a much smaller scale.

Advances in fabrication techniques utilizing the next generation of biomaterials will provide much needed insight into interfacial biology, complex tissue engineering, and vascularization of large tissues to address critical questions for the transition from the bench to bedside in regenerative medicine. When these are all addressed in tandem, the resultant biomaterial will be large enough to be compatible with critically sized defects, while at the same time will provide mechanisms for the rapid vascularization of the scaffold, which still remains one of the most challenging aspects of tissue engineering.

REFERENCES

1. Burns JW. Biology takes centre stage. *Nat Mater.* 2009;8:441–443.

2. Editorial. Boom time for biomaterials. *Nat Mater.* 2009;8:439.

3. Quarto R, Giannoni P. Bone tissue engineering: Past–present–future. *Methods Mol Biol.* 2016;1416:21–33.

4. Cross LM, Thakur A, Jalili NA, Detamore M, Gaharwar AK. Nanoengineered biomaterials for repair and regeneration of orthopedic tissue interfaces. *Acta Biomater.* 2016;17:30300–30302.

5. Zhang YS, Yue K, Aleman J, Mollazadeh-Moghaddam K, Bakht SM, Yang J et al. 3D bioprinting for tissue and organ fabrication. *Ann Biomed Eng.* 2016;28:28.

6. Mandrycky C, Wang Z, Kim K, Kim DH. 3D bioprinting for engineering complex tissues. *Biotechnol Adv*. 2016;34:422–434.

7. Gudapati H, Dey M, Ozbolat I. A comprehensive review on droplet-based bioprinting: Past, present and future. *Biomaterials*. 2016;102:20–42.

8. Ball O, Nguyen B-NB, Placone JK, Fisher JP. 3D printed vascular networks enhance viability in high-volume perfusion bioreactor. *Ann Biomed Eng*. 2016;44(12):3435–3445.

9. Lancaster MA, Knoblich JA. Organogenesis in a dish: Modeling development and disease using organoid technologies. *Science*. 2014;345:1247125–1247129.

10. Mallick KK, Cox SC. Biomaterial scaffolds for tissue engineering. *Front Biosci*. 2013;5:341–360.

11. Bose S, Vahabzadeh S, Bandyopadhyay A. Bone tissue engineering using 3D printing. *Mater Today*. 2013;16:496–504.

12. Melchiorri AJ, Hibino N, Best CA, Yi T, Lee YU, Kraynak CA et al. 3D-printed biodegradable polymeric vascular grafts. *Adv Healthcare Mater*. 2016;5:319–325.

13. Trachtenberg JE, Placone JK, Smith BT, Piard CM, Santoro M, Scott DW et al. Extrusion-based 3D printing of poly(propylene fumarate) in a full-factorial design. *ACS Biomater Sci Eng*. 2016;2(10):1771–1780.

14. Wang MO, Vorwald CE, Dreher ML, Mott EJ, Cheng MH, Cinar A et al. Evaluating 3D-printed biomaterials as scaffolds for vascularized bone tissue engineering. *Adv Mater*. 2015;27:138–144.

15. Kuo C-Y, Eranki A, Placone JK, Rhodes KR, Aranda-Espinoza H, Fernandes R et al. Development of a 3D printed, bioengineered placenta model to evaluate the role of trophoblast migration in preeclampsia. *ACS Biomater Sci Eng*. 2016;2(10):1817–1826.

16. Wheelton A, Mace J, Khan WS, Anand S. Biomaterials and fabrication to optimise scaffold properties for musculoskeletal tissue engineering. *Curr Stem Cell Res Ther*. 2016;13:13.

17. Karande TS, Ong JL, Agrawal CM. Diffusion in musculoskeletal tissue engineering scaffolds: Design issues related to porosity, permeability, architecture, and nutrient mixing. *Ann Biomed Eng*. 2004;32:1728–1743.

18. Sobral JM, Caridade SG, Sousa RA, Mano JF, Reis RL. Three-dimensional plotted scaffolds with controlled pore size gradients: Effect of scaffold geometry on mechanical performance and cell seeding efficiency. *Acta Biomater*. 2011;7:1009–1018.

19. Delplace V, Obermeyer J, Shoichet MS. Local affinity release. *ACS Nano*. 2016;10:6433–6436.

Index

Printed and bound by CPI Group (UK) Ltd, Croydon, CR0 4YY

01/11/2024

01782619-0002